ECONOMICS OF SOIL EROSION

About the Author

Pushpam Kumar, Reader at the Institute of Economic Growth, holds a Ph.D. in Environmental Economics from Delhi University. He has taught Master's level students in the Department of Business Economics, Delhi University for several years. He was the Member Secretary to the World Bank constituted Environmental Economics Training Committee for Practicing Economists and Non-Economists (EETCPN) of the 'World Bank Aided India Environmental Management Capacity Building Technical Assistance Project in Environmental Economics'. He is involved in many international collaborative research projects and currently he is Coordinator of Responses Working Group of the Millennium Ecosystem Assessment (MA). His research interests include valuation and accounting of ecosystem services, bioprospecting and corporate social responsibility. His earlier book was *Linkages of Environmental Economics and Sustainable Development: Some Issues in Modelling, Accounting and Valuation,* Manak Publications, Delhi, 2001.

ECONOMICS OF SOIL EROSION

Issues and Imperatives from India

Pushpam Kumar

Institute of Economic Growth
Delhi Univeristy Enclave, Delhi

CONCEPT PUBLISHING COMPANY, NEW DELHI-110059

ISBN : 81-8069-072-5

First Published 2004

Published and Printed by
Ashok Kumar Mittal
Concept Publishing Company
A/15-16, Commercial Block, Mohan Garden
New Delhi-110059 (INDIA)

Phone : 25351460, 25351794
Fax : 091-11-25357103
Email : publishing@conceptpub.com

Preface

Along with air, water and biological resources, soil is another crucial natural resource needed for sustenance of human well being as 97 per cent of food requirement for humans comes from the terrestrial ecosystem. With the erosion of topsoil, many civilization in the past have vanished from the face of the planet earth. Even today soil erosion and land degradation remain some of the most serious challenges, the society is confronted with. Most of the drivers of change leading to the phenomenon of soil erosion fall in the domain of economic and social factors. Yet, economists have been able to pay only a scanty attention to this problem.

This book attempts to analyse the approaches of economists to the problem of soil erosion. Besides assessing the degree and dimension of the problem in India and the World, causes and consequences of soil erosion in terms of economic implications have been mapped out. With the help of a case study of Dehradun Valley in Uttranchal State, different approaches of economics to deal with this problem has been closely scrutinized. The inference in this book would hopefully, provide crucial inputs to the decision makers and other conservation managers.

In order to make it as an easy reading, a round of revisions has been done. In this process, the current content of the book has for deviated from what it was in the beginning. If the readers find this book in the interdisciplinary tradition of ecological economics, filling the void between ecology and economics, which is absolutely essential for sustainable policies and practices, I would consider my effort as meaningful.

Pushpam Kumar
Institute of Economic Growth
University of Delhi Enclave
Delhi-110007

Acknowledgements

In order to accomplish this book that has been written in the interdisciplinary tradition of ecological economics, I have benefited from discussions with people working in different areas in last two years. During my study, a large number of people have encouraged and helped me in various ways. Citing each and every name would be a marathon task. Neither I am planning to do that nor I think, omission of any name lessens the importance of their contribution. However, I will be failing from my duty if I do not acknowledge to few of them.

First and foremost, I would like to express my sincere gratitude to Professor Dalip S. Swamy – a former Professor at the Department of Business Economics, Delhi University who was instrumental in developing my ideas and understanding of the issue. In spite of his critical health condition, he showed extraordinary interest in the study and sharpened my focus by offering penetrating comments and suggestions. I am extremely thankful to him. Dr. Arun Bhaskar and Dr. M.K. Pandit, Senior Scientists with the Centre for Inter-Disciplinary Studies of Mountain and Hill Environment (CISMHE), University of Delhi were of crucial help to me during my study. They volunteered to share their immense expertise on Doon valley with me. I am grateful to both of them. My thanks are due to the soil scientists and geologists of the Central Soil and Water Conservation Research and Training Institute, Dehradun, especially Dr. Ram Babu, Dr. G. Sastry, Dr. G.P. Juyal and Dr. Gurmel Singh who during my several visits to Dehradun spared their precious time to discuss various issues and intricacies of soil erosion in the valley, with me. They also suggested constructive direction, for my research. At the initial stage of the study, Prof. Gopal K. Kadekodi of the Institute of Economic Growth (then) helped me in preparing the skeleton of the study proposal that ultimately became the foundation stone of the present work. I am thankful to him.

During the several years of work, Prof. C.R. Babu kept encouraging me by reminding me the significance of the study which I had undertaken. He was also able to keep a vigil on my indolence. I am immensely thankful to him. At the Institute of Economic Growth, I have benefited from long stimulating discussions with Prof Kanchan Chopra on various issues of ecological economics that have sharpened my thought and analytical ability. Sometime we agreed and sometimes we disagreed. Her dissent helped me more than her consent that compelled me to rethink about the paradigm of ecological economics. I am greatly thankful to her.

Library staffs of the Institute of Economic Growth, Delhi University (South Campus); Ratan Tata Library of Delhi School of Economics; CSWCRTI, Dehradun; Indian Institute of Remote Sensing, Dehradun; Forest Survey of India, Dehradun and Forest Subdivision, Dehradun have been extremely cordial in providing me all necessary facilities. District Statistical Officer, Dehradun was of immense help to me. My sincere thanks go to all of them. My thanks are due to Shri A. K. Mittal of Concept Publishing Company, who showed sufficient interest in the manuscript and did everything to bring this book on time.

My wife Manasi always came forward with readymade suggestions and advices on anything and everything during the study period that facilitated my work directly or indirectly. I owe my special thanks to her.

Last but not the least, in my house, I was well taken care of by domestic help, Padam, a Gorkha boy, who not only regularly made my deranged study room as cozy but even awoke with me in the night to serve me the most needed tea with all smile on his face. Many thanks to him!

Pushpam Kumar

CONTENTS

LIST OF TABLES

LIST OF FIGURES

1

Introduction

All fundamental economic activities directly or indirectly depend upon inputs provided by different ecosystems. Soil, water, air and many other biological resources enter into our consumption and production processes at different stages and in different forms. Various ecological functions contribute unique economic goods and services having no closer substitute. The production and consumption activities also dump its residuals into the ecospace. In absence of these facilities of sink, the cost of treatment will be very high. Thus economic system is dependent on ecological system. Robust economic system on a continuously growing path needs a healthy and resilient ecological system. Only a healthy ecosystem can provide crucial inputs to the economic system and assimilate its wastes. Any damage to any component of the ecological system will start applying a limit on the economic growth and will inflict a heavy cost on the individual as well as society.

Soil, which is a productive layer of earth's surface with regular structure and specialized feature, is a subsystem within the ecosystem. Soil sustains the agricultural activities through its plants nurturing roles. Around 97 per cent of society's food come from the land covered with soil. Soil has many other important ecological functions but the contribution to agriculture alone makes it vital for sustenance and well-being of human civilisation. This crucial component of ecosystem is being degraded and eroded all over the globe due to our short-sightedness and unscientific use of it. Around 35-40 per cent of the earth's land are already degraded and the rate of erosion of soil is ever increasing. The present rate of soil erosion which ranges between 10 to 100 t/ha/year, is at least 10 times that of soil formation. Agriculture activities causing to erosion, salinization and water logging comes to be around 6 million hectares per year at global level.

In India around 41 per cent of the total geographical area is degraded from moderate to very severe scale (MOA, Government of India). Typical to the problem of environmental resources, the degradation of soil too, can be attributed to the phenomena of information failure, market failure and intervention failure. These failures have not got much focus in the mainstream discussion of soil management.

The research findings in soil science explicitly focus on its physical characteristics, causes and impacts of erosion and ways to halt the excessive erosion. Soil scientists have demonstrated a huge success in measurement and management of soil erosion. They have successfully evolved techniques and tools for answering questions arising out of alarming state of this resource. They look forward to technology as the panacea of the problems of erosion. Technology rightly is the answer to the problem of natural resource management, however one should raise the issue that technology may be the answer but what is the question? The question here should be the direction and dimension of technology induced welfare change of people. To answer these questions, economic and social dimension of use, misuse and management of natural resources in general and soil in particular must be analysed in scientific but interdisciplinary framework. So far most of the studies have been in the domain of pure natural science creating a hiatus between the solution of problem (natural science) and its adoption (social science). This book in the tradition of ecological economics based on transdisciplinary premises tries to move in the direction of bridging this gap and provides an economic analysis to the problem of soil erosion in India by selecting a case of Dehradun Valley popularly known as Doon Valley.

The exploration of economics of soil erosion begins with the introduction of soil as a natural resource (Chapter 2). The fundamental features of soil and its erosion process have been discussed. Accelerated soil erosion has been attributed primarily to the anthropogenic activities. In this context the responses of the economics profession to the problem of soil erosion has been mapped out for last 200 years. Soil and the related issues have been tackled by the natural scientists but soil as a useful resource has pure economic value. So the economists who have provided direction and vision to modern economics science and have been discussed and the concern for soil has been discussed briefly (Chapter 3).

In Chapter 4, an attempt has been made to take a stock of the soil erosion around the globe. Soil erosion its meaning, measurement and extent involves some amount of subjectivity, therefore, a large number of sources like international agencies (*e.g.*, the Resources for the Future), Government Department (*e.g.*, Ministry of Agriculture), Research Institutes (*e.g.*, World Watch Institute) and individual scholar's views have been incorporated. Focus remains on water induced erosion. Soil erosion takes place due to rain and wind but the economic activities through land clearing process accelerates the process. A separate chapter (Chapter 5) has been devoted to identification of factors causing soil erosion falling in the domain of economics.

If soil erosion is exceeding the rate of its natural formation then what is the effect on different economic agents (individual as well as society)? This question has been analysed in Chapter 6. The impacts have been categorised into on-site and off-site. Examples of these impacts have been carefully chosen from world and Indian scenario to highlight the issue with utmost accuracy. Chapter 7 is the fallout of Chapter 6 where various economic valuation methodology requiring to estimate the impact of soil erosion has been discussed and analysed with the help of appropriate examples from India and abroad.

Further, since the process of erosion is aggravated by several socio-economic factors inflicting a heavy cost to the individual (farmers) and society. A valley area in Dehradun (Doon Valley) of Uttaranchal state in India has been selected to examine the underlying postulates in economics of soil erosion. This area comprises of two major watersheds of Song and Asan and covers around 2.09 Lac ha of area. The valley has been experiencing water induced erosion due to unsustainable landuse change. Chapter 8 contains the general depiction of the study area, discussing its geological, biological and socio-economic features. Subsequently soil loss in the valley has been estimated using the Universal Soil Loss Equation (USLE). The loss of soil has been estimated for 25 years of time period starting from 1970 after every 2 years of time period. The estimation of soil loss shows an increasing trend and confirms to various similar studies for Doon Valley undertaken by the Central Soil & Water Conservation Research and Training Institute (CSWCRTI), Dehradun and Indian Institute of Remote Sensing (IIRS), Dehradun.

In Chapter 9 the cost of erosion has been estimated by following resource value and production value approach. As has been the general findings in all cases, the resource value exceeds the production value of soil erosion. Our exercise for Doon Valley also confirms to this. Chapter 10 deals with the conservation strategy for severe and very severe erosion category and its economic cost has been estimated. In economics of soil erosion, benefit cost analysis of conservation measures is an essential component in order to justify the economic rationale of conservation efforts. It further analyses the soil conservation practices in the valley on the basis of Benefit-cost criteria. It also discusses the reason for absence of adequate conservation efforts on the part of farmers in the valley.

Finally, based on our analysis, policy measures for soil conservation have been suggested.

2

Soil as a Natural Resource and its Erosion

Among the precious natural resources on which human's consumption and production systems rely, soil commands a special place. Soil means the top 15 centimetres thick mixture of clay, silt, organic matter, air, sand, mineral and water which provide invaluable ecosystem services functions to the mankind. Soil has been defined in various ways by the soil scientists. Here in the book, unless cited otherwise, soil refers to the loose material composed of weathered rock, other minerals and partly decayed organic matter, that covers large parts of the land surface of the Earth. Soil has got different connotations for different professions. For agriculturists, soil is that portion of earth's surface on which he can plough and grow crops. For a civil engineer, soil provides the foundation of construction activity—roads, highways, etc. For oil technologist, soil clays and clay minerals are the source of petroleum cracking agents and drilling muds. In the present text, as mentioned earlier, the thin layer of earth's crust, which serves as a natural medium for the growth of plants, will be the centrepoint of discussion.

Soil is an important component of the terrestrial ecosystem. A terrestrial ecosystem consists of trees, herbs and grasses (primary products) and decomposes like micro-organism etc. The primary producers use some of the energy from the sun to convert atmospheric carbon dioxide into organic compounds. This process reduces water and nutrients, which come from soil. Thus soil generates organic compounds, which are used by herbivores and carnivores as a source of energy. This process is the corner stone of biotic system, which provides the necessary flow of goods and services. The role played by soil in the whole process provides us with the precious ecological functions. Thus, a loss of this valuable resource will have high opportunity costs.

2.1 Soil—A Natural Resource

Soil performs significant role in the growth and development of a plant. It provides water, nutrients and mechanical support for plants and trees in natural forests and grasslands, annual and perennial crops and planted grassland. Soil serves as a natural storehouse for the plant food and can help in sustaining a plant without any fertilizers for a fairly long period of time. It also provides the habitat for decomposer organisms, which have an essential role in the cycling of carbon and mineral nutrients—the basis for economically significant ecological functions. Soil acts as a buffer for temperature change and for the flow of water between the atmosphere and groundwater through various chemical and microbial activities. It also provides the basic requirement for seed germination and hence the crucial help for managed as well as natural ecosystem.

All these functions and features of soil should not compel us to understand that these are the only roles played by the soil resource. The mankind simply depends upon soil for all its activities, whether it is agriculture, industry, infrastructure or any recreational purpose. But it is agriculture through the plant sustaining capability of soils, which matters the most for our survival and prosperity. More than 97 per cent of the world's food comes from the land rather than the ocean and other aquatic systems (Pimental, 1993). Cultivation of soil to grow grains like wheat, barley and other crops originated about 12000 B.C. in the valleys of northern upland area of Mesopotamia, now Iraq (Wild, 1993). Over a period all civilisations, their prosperity and doom have been closely linked with the varying degree of soil fertility. Many flourishing civilisations probably collapsed due to the declining soil fertility (Jack and White, undated). Even today only those countries who use soil judiciously are likely to sustain economic growth.

It is clear that all primary needs like food, fibre, fuel etc., come from soil and their ecological functions. Therefore, soil needs special attention. Today 6 billion humans exist on this earth and additional 2,50,000 humans are born each day that have to be fed besides the need of the bovine population.

2.2 Erosion of Soil

The word 'erosion' is of Latin origin derived from the verb 'erred'—to eat away or to excavate. The term soil erosion was introduced in the beginning of the 20th century (Zachar, 1982). Bennet *et al.* (1955) considers that soil erosion in the true sense of the word, should only refer to that erosion which is more severe in intensity than the erosion that takes place in natural plant communities undisturbed by man. Going by this view even a mild degree of erosion, activated by humans will be considered as erosion but severe erosion done by wind will not be.

This problem has been eliminated by introducing the scheme in which, erosion is replaced by the word denudation done by man (man-accelerated soil erosion) or normal denudation. In describing the total removal of silt by rivers one can speak only of denudation. Consequently, in the broad concept of denudation as a removal process, soil erosion can be considered, in the more narrow sense of the word, to be only that erosion (denudation) which is accelerated by man (Zachar, 1982).

By all accounts it will be safe to infer that the categorization of erosion into normal, natural, accelerated etc. is not satisfactory. Many phenomena occurring in nature is not 'normal' which cause the existence of normal soil. This means that Natural Soil, *i.e.* soils which has developed without human interference may, however, occur in areas where 'normal' soil profile does not exist. Here soil erosion can not be depicted through the term accelerating erosion, for the reason that humans may not only accelerate erosion but he may halt the process of erosion, too. Hence de-accelerated erosion must also be considered as soil erosion. Many experts opine that erosion has occurred only since and where man has been engaged in agricultural activities. This is also not acceptable because soils were affected long before artificial changes began to be made.

The earlier concepts although credible, fail to highlight the complexity of the issue and at times, are aberrant from reality. Broadly, soil erosion can be put to dichotomous classification of natural and artificial (anthropogenic). Natural erosion can further be split into two broad categories, *i.e.* normal and abnormal. Normal erosion refers to the erosion under normal conditions leading to the formation of 'normal' soil. In the case of abnormal erosion, removal is so great that normal profile of soil fails to develop. Similarly, artificial erosion can also be divided into accelerating and decelerating. Reduced erosion under control process and conservation practices is known as decelerated erosion. In accelerated erosion, main interference causes excessive erosion over and above the normal one.

The process of soil formation takes place along with and parallel to the process of erosion. The rate of normal erosion may exceed the rate of formation of soil or may remain below this critical limit. Similarly, altered erosion rate may cross the limit of formation rate. If erosion rate is lower than that of formation, overall impact is benignant. Otherwise, it becomes malignant in character.

Soil erosion, whether natural or altered, becomes significant and needs attention when it exceeds the critical limit of soil formation.

2.3 Desertification, Land Degradation and Soil Erosion

In common parlance, terms like desertification, land degradation and soil erosion are frequently used synonymously. But they are different concepts having different bases and entirely distinct interpretations. These features are most common in semi-arid and dry sub-humid environment, but can also be found in some tropical and subtropical humid and temperate environments in both Less Developed Countries (LDCs) and Developing Countries (DCs) (Fantech & Margris, 1986). There are various interpretations and definitions of desertifications. The UN Conference (1977) on desertification (Resolution No. 7) defines it as the denudation or destruction of the biological potential of the land, and leading ultimately to desert like conditions. It is an aspect of the widespread deterioration of ecosystems and has diminished or destroyed the biological potential, *i.e.* plant and animal production. What matters ultimately in this key concept of desertification is the outlook of various groups of people involved. There may be enough indicators, which go, unnoticed or unheeded (Heathcote, 1980). Land degradation on the other hand means any undesirable alterations in the form of the land in use from the cultivation and other economic use point of view. Land degradation is an umbrella concept encompassing soil salinity, water logging, and nutrient loss and soil erosion.

Obviously, desertification, which is a kind of land degradation in arid, semi-arid and dry sub-humid areas resulting from various factors including climate variations and human activities, falls under this umbrella term. All types of deformations, pollution and alteration in the normal colour, texture, composition, shape, fertility, mobility and organic variety and variability of the ecosystem of soil can be called as land degradation. At times, land degradation may be referred to in context of some cultivable land or cropland. However, soil erosion remains the dominant form of land degradation. In fact around 84 per cent degradation is attributed to the problem of soil erosion. The remaining 16 per cent degradation consists of all other kinds of degradation.

As we move further, it is the soil erosion (especially due to water erosion), which takes the centre space albeit other types of land degradation have also been addressed adequately.

3

Evolution of Approaches and Responses of Economists to the Problem of Soil Erosion

3.1 Introduction

Soil is one of the fundamental natural resources. Growing demand for food through cultivation crucially depends upon this resource. In recent years, this precious natural resource has been dwindling because of its erosion. The mankind simply depends upon soil for all its activities, whether it is agriculture, industry, infrastructure or any recreational purpose. Over a period all civilizations their prosperity and doom have been closely linked with the varying degree of soil fertility. Many flourishing civilizations probably collapsed due to the declining soil fertility (Jack and White, undated).

Soil erosion are caused by natural factors (*e.g.* rain, wind) and anthropocentric factors (*e.g.* cultivation on hills, long fallow period and other similar practices of farming etc.). Soil erosion whether it is natural or man-made becomes serious when the erosion rate exceeds the critical limit of its formation. The Global Assessment of Soil Degradation (GLASOD) estimates that over the period 1950-90 about 17 per cent of the world's entire vegetated areas lost some of its soil productivity (FAO, 1994). In 1991, a UN study estimated that 552 million hectares (equal to 38 per cent of today's global cultivated area) had been damaged to varied extent by agricultural mismanagement since the Second World War (Gardner, 1997).

Soil erosion has primarily been discussed and researched by the natural scientists (*e.g.*, geologists and engineers). Soil erosion may be a natural phenomena but it has got serious economic implications in terms of its impact on people's welfare in the society. In the mainstream literature on soil erosion, economic dimension of the problem of soil erosion has either been missing or has been dealt perfunctorily. Erosion of top soil is primarily caused by anthropocentric activities and it can either be halted or be reduced by altering the behaviour of the soil users, *e.g.* farmers. And hence the role of economists becomes very important.

Conventional economists have not been explicit and consistent on the issue of soil erosion. In the 18th and 19th centuries, land along with capital and labour were recognised as fundamental factors of production. It would be worthwhile to mention it here that land and soil were interchangeably used in the literature during this period. During 18th and 19th century land remained the central point of discussion when classical political economists talked about natural resources especially scarcity, abundance of land and future feasibility of continuing path of economic growth and well-being of the society. In most of their works, the issues of land quality, their varying fertility and appropriate conservational efforts etc. are sufficiently represented. In fact classical economists were more concerned and disturbed about the issue of land degradations during 18th and 19th century than their counterparts of 20th century. The latter were so preoccupied with the ideas of technological revolution and efficiency issue that the fundamental issue of land resource got marginalised.

In order to analyse the philosophical underpinnings behind the formation of response of dominant economists to the problem of soil erosion or land degradation, and their perception of value in the context of soil erosion, three major categories, *e.g.* classical, neoclassical, interventionists and others and other groups have been identified. Their response to the problem of soil erosion has been analysed and explored in the following sections.

3.2 Classical Political Economists

The approach and responses of classical political economists to the problem of soil erosion and land degradation does not seem to be coherent and homogeneous. Each economist speaks according to his socio-political environment. The words—soil erosion, land degradation and soil degradation — have been used interchangeably by these economists. Prominent voices among them were of Adam Smith, David Ricardo, J.S. Mill and Karl Marx. Starting with Adam

Smith, one can easily infer this. Through the publication of *The Wealth of Nations*, Adam Smith emphasised on the phenomenon of self-correctedness and invisible hand. He was a product of the time when agriculture and industry were developing rapidly. He did not see any constraint on the availability of land resources. Smith did discuss the idea of rent, but did not distinguish between rent from food producing land and mineral producing land (Kula, 1998).

Adam Smith is essentially highly optimistic about the adequacy of natural resources; especially agricultural land. During his time, a few technological developments (although modest in nature) perpetuated his feelings about the abundance of the arable land. Environmental problems and depletion of natural resources were not noticeable during his time. As for the crude produce of the land, such as useful deposits, precious stones, energy resources, mineral, cotton and so forth, available facts at the time did not undermine his optimism about future prospects. Adam Smith was more interested in the cause of freedom of economic growth which fitted his rationale. He thought that state should provide all necessary conditions so that the rational human being acting in his self-interest shall pursue his independent course of actions and this process shall ensure future growth and prosperity without harming the environment inclusive of natural resources.

Adam Smith's arguments could not sustain the wrath of the socialists of the 19th century. They badly thwarted the idea of free interplay of private interests in the society. Today, our world is very different from that one in which Smith lived and many of us are becoming increasingly concerned that the doctrine of *laissez faire* could lead us to an environmental catastrophe. There is ample evidence in the form of acid rain, the greenhouse effect, ozone depletion, deforestation, land and water contamination, soil erosion and accumulation of vast quantities of highly toxic nuclear or other wastes, all of which are created and aggravated by the pursuit of narrow interests (Kula, 1997). Obviously due to his priorities and contemporary perceptions, Adam Smith could not add anything significant to the problem of soil erosion which commands very serious attention today.

Thomas Malthus on the other hand, with the help of his book — *An Essay on the Principle of Population as it Affects the Future Development of Society* (1798) was a contrast to the optimist Adam Smith. Malthus carefully observed the growth of population in some northern states of America and came to the conclusion that population had the tendency to double itself every 25 years period of time. He gave the idea that the population growth followed the path of geometric progression. On the other hand, food supply follows the path of an arithmetic progression. Malthus in his Essay states that:

> " ... Man is necessarily confined in a room. When acre has been added to acre till all the fertile land is occupied, the yearly increase of food must depend upon the amelioration of land already in possession. This is a fund, which form the nature of all soils, instead of increasing must be gradually diminishing..."

Malthusian idea of natural resource scarcity emanated from the fixed arable land. He visualised a bleak future for mankind due to the widening gap between the rate of food supply and population growth. Whatever assumptions compelled Malthus to prophesize doom for mankind, his opinion on land resources did not have much empirical basis. But today it has evolved into a viable proposition because of alarming state of global cropland. Growing land degradation is primarily an affect of population explosion. During his life-time and even after his work as a light house his views have cautioned the mankind about its future. He influenced most of his contemporary thinkers and philosophers including biologists like Charles Darwin.

David Ricardo possibly, was the only economist who systematically discussed about the land resources (arable land) and made it the basis of his famous theory of rent. Ricardo had strong bearings of Malthusian theory but his theory related to land was fundamentally unique and distinct. Ricardian scheme clearly demonstrated that economic growth must sooner or later peter out owing to the scarcity of natural resources. In 1817, Ricardo published *Principle of Political Economy and Taxation* in which he wrote that:

> "Rent, it must be remembered, is not in proposition to the absolute fertility of the land in cultivation, but in proposition in its relative fertility. Whatever cause may drive capital to inferior land must elevate rent on the superior land; the cause of rent being, as stated by Mr. Malthus in his third proposition, the comparative scarcity of the most fertile land. The price of corn will naturally rise with the difficulty of producing the last

> portion of it, and the value of the whole quantity produced on a particular farm will be increased, although its quantity will be diminished; but as the cost of production will not increase on the more fertile land, as wages and profits taken together will continue always for the same values, it is evident that the excess of price above the cost of production or in other words, rent must rise with the diminished fertility of the land, unless it is counteracted by a great reduction of capital, population and demand" (Ricardo, 1817).

Ricardo explicitly recognised the different levels of soil fertility which gave rise to the concept of rent. Ricardo also stated that the shortage of natural resources will limit the growth process (arable land). But in his scheme, scarcity of arable land got reflected through rising market prices. Hence, he talks about relative scarcity of natural resources. Ricardo defines rent as that portion of produce of the earth which is paid to the land owners for the use of the original and indestructible powers of the soil. Ricardo faced a formidable challenge from various corners but Henry Caney's (1793-1873) was most dominant one and he also proved the Ricardian paradigm as superficial on empirical basis.

John Stuart Mill elaborated on the ideas of Malthus and Ricardo but in a wider perspective. In his *Principles of Political Economy* (1948, ch.12) he wrote:

> "Land differs from the other elements of production. Labour and capital in not being susceptible of indefinite increase. Its extent is limited. It is also evident that the quantity of produce capable of being raised on any given piece of land is not indefinite. This limited quantity of land, and limited productiveness of it, are the real limits to the increase of production" (Mill, 1948).

Mill, however, did not agree with the absolute limit of resources in Malthusian schemes. Mill acknowledged the great potentialities of technology and institutions. Mill, possibly was much ahead of his time because he not only recognised the importance of arable land and their important roles in the growth process, he also clearly gave due importance to the virgin land and associated wilderness. He states:

> "A population may be too crowded, though all be amply supplied with food and railment. It is not good for man to be kept perforce at all times in the presence of his species. A world from which solitude is extirpated is a very poor ideal. Solitude, in the sense of being often alone, is essential to any depth of mediation or of character; and solitude in the presence of natural beauty and grandeur is the cradle of thoughts and aspirations which are not only good for the individual but which society could do without it. Nor is there much satisfaction in contemplating the world with nothing left to the spontaneous activity of nature; with every road of land brought into cultivation, which is capable of growing food for human beings; every flowery waste of natural posture ploughed up, all animals or birds which are not domesticated for man's use exterminated as his rivals for foods, every hedgerow or superfluous tree rooted out, and scarcely a place left where a wild shrub or flower could grow without being eradicated as a weed in the name of improved agriculture" (Mill, 1948).

Mill tried to elaborate on the existing classical doctrine of resource scarcity and land degradation. In his elaboration upon the ideas of Malthus and Ricardo, as Barnet and Morse (1965) had angled, Mill took account of, and lent credence to, nearly all the qualifications of scarcity doctrine that seem relevant today.

Karl Marx had tremendous impact on the political, economic and sociological thoughts emerging in the 19th and 20th century*. As Marx in his attempt to provide a systematic account of development, is not bothered about the scarcity of natural resources. He did write about land resources and their importance but his views were at the most perfunctory. Marx was busy in developing the idea of labour value of surplus and attributes human labour as the ultimate resource. He wrote —

> "The land (and this, economically speaking, includes water) in its original state in which it supplies man with necessaries or

* See Blaug (1997) who writes, "Marx has been reassessed, revised, refuted and buried a thousand times but he refuses to be relegated to intellectual history. For better or the worse, his ideas have become part of the climate of opinion within which we all think. No one now does battle for Adam Smith or Ricardo but blood pressure still go up when Marx is under examination."

> means of subsistence ready to hand is available without any effort on his part as the universal material for human labour. All those things which labour merely separates from immediate connection with their environment are objects of labour spontaneously provided by nature, such as fish caught and separated from their natural—namely, water, timber felled in virgin forests and ores extracted from their veins. If on the other hand, the object of labour has been filtered through previous labour, we can call it raw material. For example, ore already extracted and ready for washing. All raw material is an object of labour but not every object of labour is raw material; the object of labour count as raw material only when it has already undergone some alteration by means of labour... which the exception of the extractive industries such as mining, hunting, fishing (and agriculture, but only in so far as it starts by breaking up virgin soils), where the material for labour is provided directly by nature, all branches of industry deal with raw material, *i.e.* an object of labour which has already been filtered through labour, which is itself a product of nature" (Marx, 1867).

Marx considered the value of product in the context of a social relationship that arise historically and are specific to certain societies or mode of production. Natural resources such as land, which Marx regarded as free gift of nature, on which no human labour has been extended has substantial exchange value. In Marxian thought, there are four possible theories of rent—monopoly theory, differential rent theory, capital theory of rent and Marx's own theory. In the first, rent comes about from the monopoly price of agricultural commodities and the monopoly price of the land ownership. The price of agricultural produce is higher than its value because of its supply being below the demand. This arises due to the law of diminishing returns and thus the supply cannot keep up with demand. The second theory coincides with the first one in that differentials in rent are due to the variations in land fertility. The third relates to the use of capital on land in order to improve its fertility. Here, rent is linked to the interest on capital equipment used to enhance productivity. However, this theory is incapable of explaining the rent on land on which no capital has been used (Kula, 1998).

Marx's work *Das Kapital* although had sufficient remarks about land resources and its degradation but more elaborate treatment could be found in his another work, *Theories of Surplus Value* (1859, p. 69). Here, Marx did not distinguish among different kinds of natural resources, *i.e.* fisheries, forestry, land etc. and he held capitalist class and their capitalist mode of production responsible for environmental depletion and degradation. Further, Marx, in line with Adam Smith ruled out the possibility of the rule of diminishing returns on land resources as propounded by Malthus and Ricardo. The number of reasons forwarded by Marx for this were multifold. First, progress in science and technology would outplace physical difficulties encountered in natural resource based sectors, including farming. Poor soil can become viable as agricultural technology improves. Second, improved communication methods may bring better soil as well as richer mine deposits, into use. Third, institutional changes could expand production; for example, the release of previously unworked communal land might be put into use.

Marx's writing on natural resources and their scarcity were definitely not dogmatic. He explained human labour as the ultimate resource but he acknowledged the role of nature as well. In the critique of the Gotha Programme Marx argued that "labour is not the source of all wealth, nature is just as much the source of use values (and surely there are what make up material wealth!) as labour. Labour in itself is only the manifestation of a force of nature, human labour power." One cannot ignore the fact that there exists a possibility where degradation and depletion of natural resources could have been explained beyond the boundary of social relations but unfortunately Marx either could not think on that line or his prime concern for worsening plight of industrial workers and his penchant for 'class struggle' theories did not allow him to do so (Gowdy, 1984; Barbier, 1989).

For the major part of his creative life, Marx remained attached to the woes of capitalist system where he attributed each and every ill of the economic system or ecological system to the capitalist mode of production. Marx however had identified capitalist agriculture as the main reason for exhaustion of arable land:

> "According to Marx the premature exhaustion of the soil occurs because there are economic disincentives to maintain long-lived capital investments. With increasing resource exhaustion, the

> rising cost of raw material, depress the rate of profit. Furthermore, the fact that the 'marginal cost' of a non-renewable resource is in reality only the marginal cost of extraction and not production creates a surplus for the producer. This surplus arises from the fact that the marginal social cost of producing a unit of an exhaustible resource is much greater than the private cost. Marx faults the price system for misallocating natural resource overtime. The time horizon for non-renewable and semi-renewable resources is not compatible with the time horizon for the prices the market system imposes on these resource" (Gowdy, 1984).

Marx often attributed the issue of land degradation to the ownership pattern. He prefered the collective ownership to private ownership of land to avoid its erosion and deterioration in fertility.

3.3 Neoclassical Economists

Neo-classical economists did not pay much attention to the problem of soil erosion and land degradation except some individual 'adventure' who were either greeted with casual encouragement or were branded as obscure or too scientific to be included in a social science like economics by the more influential voice in the profession. Among them were the likes of Alfred Marshall and Hotelling. Although. Menger, Javons and Walras had already laid the foundation of utilitarian framework for demand analysis and, it was Marshall to whom the credit goes for reconciliation of the demand and supply side economics. Marshall's view on agriculture was more in the line with the classical school than the tradition of his time (Shumpeter, 1954). Marshall maintained that if properly managed, agriculture can remain productive for-ever. Marshall also conceptualised the issue of externality which was later elaborated by A.C. Pigou (1920).

H. Hotelling also commands a special place here who pioneered in the analysis of renewable and non-renewable resources. Hotelling's forte lied in his capability of mathematical interpretation of exhaustive resources (Hotelling, 1931). Top soil, being an exhaustible resource for the relevant time horizon could very well fit into his scheme. But primarily, his theory was aimed at mining sector. Hotelling's analysis had been combined with characterization of the optimal path of economic growth based on Ramsey's approach which determine the pattern of investment in reproducible capital in accordance with a utilitarian social welfare function (Ramsey, 1928). Hotelling demonstrated that under optimal conditions, the price of an exhaustible resource net of extraction cost must be rising at a rate equal to the rate of interest on other assets. Thus for Hotelling, depletion of exhaustible resources was not bad if it fulfilled the above conditions. As it is a proven fact by now that top soil's productive, assimilative and maintenance roles are so profound and crucial that Hotelling's rule would lead us to an erroneous goal where we will not be able to check the process of rapidly depreciating soil resources.

3.4 The Case of Interventionist School and Land Degradation

Since the beginning of the 20th century a large number of economists started focusing on the role of state and planning process. They got convinced by the phenomenon of market failure which caused land degradation and erosion of soil resources. They believed that an effective collective body like state is better equipped to handle the issues of depletion of environmental quantity (soil, forestry etc.) and degradation of environmental quality (air, water pollution etc.). Their concern was primarily for industrial pollution through water and air. Economists like A.C. Pigou, R. Coase, K.E. Boulding and E.J. Mishan were the front runners among them.

In promotion of social welfare, Pigou expressed great concern for depleting stock of natural resources. Pigou attributed this phenomenon to our defective telescopic faculty which always prefers present to future. Here Pigou, possibly was among the first few who expounded the theory of inter-temporal externality. Pigou observed that:

> "... Sometimes people will win what they require by methods that destroy as against the future, much more than they themselves obtain. Over-hasty exploitation of the best coal seems by methods that cover up and render unworkable for over worse, but still valuable, seems; fishing operations so conducted as to disregard breeding seasons, thus threatening certain species of fish with extinctions, forming operations so conducted as to exhaust the fertility of the soil, or all instance in point" (Pigou, 1920).

Pigou made it clear that conventional market forces and their tools cannot take care of these huge negative externalities hence state interventions were necessary and will help in maximising the societal welfare. Pigou suggested basically three types of policy instruments for the rational use of exhaustible resources, protection of the environmental quality, curbing wasteful consumption, promotion of saving and above all prevention of harmful shortsightedness. These were in the form of state subsidies, taxation and legislation.

Pigou's ideas and policy formulation had a far reaching impact on the theory of environmental policy. His proposed fiscal tools became the backbone of economic instruments for abatement of pollution, adopted by many modern industrialised countries in 1960s,1970s,1980s and 1990s. However, it took time in realising that Pigouvian charges or tax imposed imperfections in the market (Turvey, 1961). Beliefs in the market mechanism denounced the Piguvian prescriptions and emphasised on consolidating the apparatus of the institute called market. Moreover, Pigouvian tax did not address the problem of soil erosion directly although one can smell its presence in various recommendations for conservation of agricultural resource base.

Instead of state intervention through charges or subsidies, Ronald Coase proposed that property rights should be ascertained and once property rights are defined then under zero transaction costs, it is possible for polluter and pollutees to bargain and arrive at some mutually acceptable and feasible solution if the magnitude of the damage inflicted on the pollutees can be evaluated properly (Coase, 1960). Coasian solution is really heroic but equally heroic are his assumptions especially the transaction costs being as zero etc. His prescriptions were path breaking but away from the problem of soil degradation.

Influential voice of Boulding cautioned the growth maniac economists about the future of natural resources and ultimately foundation of our productive economic system. Boulding's idea of 'econosphere' explicitly recognised the myopic vision of traditional circular flow of economic activity in which natural resource scarcity was hardly mentioned (Boulding, 1966, 1969). He recommended that the natural resource base must be diligently protected and wastes must be properly managed; he added that our long-term survival depends on our ability to manage these resources.

Through his works — *The Costs of Economic Growth* (1967) and

The Economic Growth Debate (1977), Mishan acknowledged the fact that the widespread public beliefs about industrial progress being ultimately beneficial, and that often economists tended to hide behind this faith when they advocated more and more growth (Kula, 1998). Mishan provided a comprehensive picture of environmental damage and contended that numerous desirable things which were freely available in the past, but have now become scarce and which, with current trends will be even more scarce in future. Mishan is of the opinion that the damage to the environment has become so great that broadening the concept of property rights to internalise external affects is no longer possible.

3.5 Other Approaches to the Problem of Soil Erosion/Degradation

Parallel to the economic approaches to land degradation, non-economic approaches (primarily of conservationists and ecologists) also became popular. Earlier they included the thinking of the contemporary economists. Pioneer among them were the French thinkers in the 18th century known as Physiocrats thinkers. Among them Quesnay and Turgot were the most famous. Defying dominant mercentalists view, they declared that it was land which created the wealth and not the precious metals and volume of trade. Quesnay published his famous book *Tableau Economighe* in 1758 and convincingly hypothesized that surplus was first created by agriculture which was then made available to other sectors and social classes. Others like Cantillon (1755) defined land as the source of all wealth and labour as the power which produced it. All material goods and their value essentially stemmed from land and labour. Physiocrats philosophers tried to provide the necessary conscience to the economist but probably lacked quantification and the hangover of mercantilist thought impaired the effectiveness of their message.

By the turn of the 20th century, views expressed by Malthus, Ricardo, Mill and Javons had already caught the public imagination, but some natural scientists did not agree much with these popular perceptions of the economists and tried to project some alternate thinking about the land and its productive capability. The whole new thinking were called as American Conservation Movement (1890-1920). Here, scientists like Marsh (1865) argued that natural environment is one of great diversity, interdependence and complexity which could not be reduced into inputs and outputs; nor can the role of man, as part of the living environment, be simplified as that of producer and consumer. Marsh was one of the key members of conservationist movement which was targeted primarily to the U.S. This conservation movement argued that the scarcity of natural resources is a fact of life; it can reveal itself as a regional or even global physical availability at a given point of time or over time (Kula, 1998). Other members of this conservation movement like Henry Thoreau, John Muir and George Santanya laid the foundation of modern environment movement all over the globe and deeply influenced the fellow economists of their time.

In 1952, *the US Presidential Commission on Material Policy* published its report entitled *'Resources for Freedom, Foundation and Growth and Scarcity'* which indicated the limited arable land in the US and its implications in the future. Another landmark in the direction of study of natural resources scarcity was a book entitled, "*Trends in Natural Resource Commodities — Statistics of Prices Output, Consumption, Foreign Trade and Employment in The United States 1870-1957*" (1962) which mentioned that during the period the population increased four-fold, land used for agriculture almost tripled, wood output increased more than three-fold. This was one of the most systematic statistical studies indicating worsening situation of land resources besides other problems.

In 1970 economists, ecologists, engineers, civil servants and philosophers expressed their concern about the factors which would limit the future rate of economic growth. Their model was heavily based on Forrestor's (1970) computerised modelling which was subjective in experience and vulnerable to many criticisms. This was known as *The Club of Rome* which propounded the Limits to Growth theory (Meadows *et al.* 1972). This Club considered five fundamental factors which had the potential to thwart the growth process. These were, namely population growth, agriculture activity, natural resource availability, industrial activity and pollution. They expressed their concern about the continuously degrading global arable land. In its attempt to formulate a Global Agenda for change, the World Commission on Environment and Development (WCED) explicitly recognised the problem of soil and land degradation. The Commission, in 1987 declared that — soil degradation eroded the overall resource base for agriculture. The loss of cropland encouraged farmers to overuse the remaining land and to move into forests and into rangelands. Sustainable agriculture cannot be based on methods that mine and deplete the soil (WCED, 1987, p.126). Commission takes a global look of the problem and attempts to provide the accurate dimension of the soil erosion which it puts as 25-30 per cent of the global total land under cultivation.

In 1992, the United Nations Conference on Environment and Development (Rio) also raised concern about the issue of soil erosion and land degradation. In its Agenda 21 this Conference stated:

> "Land is a finite resource, while the natural resources it supports can carry over time and according to management conditions and uses. Expanding human requirements and economic activities are placing over increasing pressures on land resources creating competition and conflicts and resulting in sub-optimal use or both land and land resources. If, in the future, human requirements are to be met in a sustainable manner, it is now essential to resolve these conflicts and move towards more effective and efficient use of land and its natural resources" (Agenda 21, 1992, p.105).

The Rio Conference, in a way, is the crystallised form of Brundtland Commission Report of 1987.

These reports may not use the conventional techniques of economics to deal with the problem of soil erosion but it definitely contributed in disseminating the issues to the wider masses and decision makers which in turn affects economists as well. By the middle of the 1990s, soil erosion had already presumably established itself as a formidable challenges for economists and ecologists alike. These non-economic reports and movements made the issue of soil erosion so serious and provoking that academic discipline of economics science started taking it with utmost attention and sanctity.

3.6 What Emerges Finally?

The economists of 18th and 19th century did not address the issue of soil erosion and land degradation adequately. During their time, industrial revolution had just begun and demand for food was met by traditional cultivation pattern which did not erode the top soil excessively. Traditional agriculture had enough flexibility to incorporate the conventional wisdom of farmers who adopted proper conservational practices and kept proper fallow time and to a large extent avoided monocropping. All these were possible because population explosion was yet to occur and food security was well within control.

Adam Smith was a typical product of this context and he was optimist about the future of mankind. Adam Smith could translate a few technological revolution of his time into general conclusion that land resources will never be a problem. Today's problem of widespread soil loss and subsequent land degradation definitely abandons Smith's view and his optimism. Malthus, Ricardo and Mill forecast the doomsday of mankind on the basis of their diminishing returns rule for land resources. However, Malthus mentions about absolute limit of arable land while Ricardo emphasises relative scarcity which will get reflected through the rising price. Malthusian time-horizon was hazy and Ricardo's belief of roaring prices as a result of relative scarcity of fertile land proved as wrong. Mill's arguments were too philosophical to be translated into simple economic analysis, at least during his lifetime.

Malthus and Ricardo both converge on the scarcity of land resources but in a different way. It is easier to be convinced by Ricardo's arguments of diminishing returns because physical properties of the land and their fertility differ. The best land will be cultivated first and the poorest in the last. Hence, there will be increasing costs, *i.e.* diminishing returns to labour and capital. In Malthusian framework, quality of arable land will be in short supply after sometime — but the time horizon was never defined by him, though he does talk about absolute limit to the growth. Barnet and Moarse (1963) were among the first few who emphasized the difference between the Malthusian and Ricardian scarcity of natural resources in terms of absolute and relative frameworks.

Malthus focused upon the extensive limits of the earth, and found resource scarcity inherent in the finiteness of the globe. Ricardo emphasised the differential fertility of the individual parcels of land; and assuming that the better lands would be used first, he found declining quality to be the cause of increasing resource scarcity. Ricardo, on the other hand, believed that some unexplored land pieces always exist needing more doses of capital and labour for the same land level of output. A virtually limit applies to both but the time horizon differs. Ricardian scarcity appears long before Malthusian even if it is assumed that the homogeneous Malthusian land is of the same quality as the poorest Ricardian land to be ever brought into use.

Karl Marx gave an entirely different colour to the problem of resource scarcity, especially soil erosion. In classical view, prices of the goods (*i.e.* exchange value) are taken in terms of labour input. In neo-classical paradigms it is relative price. Marx on the other hand endorses the value in terms of social relationship which are historically determined. In last 150 years when Marxian thinking started to precipitate, land degradation has reached the alarming heights. Soil erosion has accentuated and has engulfed the major portion of the global cropland. Erosion of top — soil due to mismanagement of wrong cropping pattern has occurred more or less everywhere irrespective of their political system. Marxian optimism about the good health of natural environment in a classless society seems to be either far away from the reality or an event to take place in far distant future. Operationalisation of socialism in former USSR could not take care of land degradation. Article 18 of the last Soviet constitution stated that all necessary steps would be taken to protect and make rational use of the country's land. Soviet economy was no exception as far as soil degradation in particular and environmental problem in general was concerned. *Pearce and Turner* (1990) argue rightly that the Soviet territory was fragile and other planning blunders quickly broke its back. Only about 10 per cent of the country was suitable for lucrative agriculture and only 1 per cent of this received annual rainfall of 70 cm or more. The bulk of Soviet agriculture was conducted on arid and semi-arid lands supported by massive irrigation programmes which were fundamentally flawed. There are other socialist countries where environmental problem is no different from other capitalist countries as far the magnitude of the general features of the problem is concerned. Marxian interpretation throws some light on these problems but it definitely does not explain the objective reason and its context.

Marshall, Hotelling and Pigou make the problem more vivid for their fellow professionals. They developed a framework where economic analysis of soil erosion seems to be feasible. They also paved the way for interventionists like Pigou, Mishan, Coase and Boulding who advocated for the strong role of state. Ronald Coase's position somehow is different who does not want state to do more than allocating the property rights for common degrading land resources. It is interesting to observe that the main culprit of land degradation is the phenomena of 'market failure' but these interventionists along with others seek remedy in market-based instruments. It is also noteworthy that these interventionist's arguments were primarily focused to the problem of pollution and industrial smokes.

Development of the disciplines of ecology in the latter half of 19th century and subsequent American Conservation movement explicitly paid attention to the problem of soil erosion. During 1940s, 1950s and 1960s there was a great leap forward in soil science. Technology and imperatives for measurement of soil loss got crystallised. This development got proper place in US Presidential Commission on Material Policy, Club of Rome, and Conferences on Environment and Development in 1987 and 1992 among others. Published reports of these efforts addressed the issue of soil erosion explicitly. They also advocated that for soil conservation economic arguments are needed and it has wider appeal.

Thus we see that the evolution of responses of economists to the problem of soil erosion in last 300 years has gradually accentuated. It also seems that the approaches of economists to the problem of soil erosion have not been coherent, focused and consistent. This has partly been responsible for a hiatus between economic theory and policy formulation in the realm conservation and management of soil resources.

4

Soil Erosion in the Contemporary World

Soil erosion is not a new phenomenon. In fact, it is as old as the existence of the earth itself. Human induced activity has been causing the erosion, which exceeds its formation rate. Whenever in human history, disharmony and maladjustment between society and environment has taken place, rate of soil erosion has exceeded the rate of its formation. Man induced erosion is globally pervasive. In the past, this induced erosion had been one of the most potent factors causing the downfall of civilisations and empires. Their ruined cities now lie buried as barren wastes, which were once the world's most fertile lands (Jack & White, undated).

The exact dimension and quantifications of soil erosion has become one of the most controversial issues in recent times. Erosion is generally estimated from soil maps. These maps are prepared by different local agencies operating in different countries. Their convention, assumptions and approaches vary widely across the regions. This creates heterogeneity in measurement, preventing the scope of a meaningful comparison. However, some of the recent approaches of the Food and Agriculture Organisation of the United Nations (UN) have taken this task and they have identified other forms of soil degradation (FAO, 1976, 1982, 1991, 1992, 1994).

An effort has been made by the United Nations Environment Programme (UNEP) to map the severity of soil degradation on a global scale as the world map of the status of human-induced soil degradation. The International Soil Research and Information Centre (ISRIC) have conducted this international project, known as Global Assessment of Soil Degradation (GLASOD). The GLASOD measures all kinds of soil degradation relative to their own scale of severity. Types of degradation considered by the GLASOD are as follows:

(a) Water erosion—loss of topsoil, terrain deformation.
(b) Wind erosion—loss of topsoil, terrain deformation, overflowing.
(c) Soil Fertility decline—loss of nutrients/organic matter.
(d) Water logging—acidifications, compaction and sealing.
(e) Salinization.
(f) Lowering of the water table.

For measuring severity, the GLASOD classifies degradation into four major categories:

(i) *Light*—terrain's agricultural suitability has declined but is suitable for use in local farming systems. Restoration to full productivity is possible by modifications of the management system; organisation functions are still largely intact.
(ii) *Moderate*—terrain is agriculturally productive. Productivity is greatly reduced but still suitable for use in local farming systems. Major improvements are required to restore productivity. Organic biotic functions are partially destroyed.
(iii) *Strong*—terrain becomes non-reclaimable at farm level. Major engineering works are required for restoration. Organic biotic functions are largely destroyed.
(iv) *Extreme*—The terrain is unredeemable and beyond restoration. Original biotic functions are fully destroyed

Here it must be mentioned that light degradation may not be visible but conventional wisdom and experience of the farmers acknowledge it. In case of strong and extreme degradations, lands are generally abandoned.

4.1 Global Soil Degradation Scenario

The GLASOD estimates that over the period 1950-90 about 17 per cent of the world's entire vegetated areas lost some of its soil productivity. Out of this, 6 per cent is light; 8 per cent degradation is moderate; and 3 per cent is either severe or extreme. World's total land area is considered to be as 14,048,300,000 hectares. Total vegetated area or domesticated area, as a percentage of total land was 37 per cent in 1994. Following table provides the overall global land-use pattern.

Table 4.1: Global Land-use Pattern

Land Area (000 hectare)	14,048,300
Domestic Land as per cent of total land area in 1994*	37
Cropland (1992-94) (000 ha)	1,465,814
Percentage change since 1992-94	2.0
Permanent pasture (1992-94) (000 ha)	3,410,203
Percentage change since (1992-94)	3.2
Forest and Woodland (1992-94) (000 ha)	4,177,088
Percentage change since (1992-94)	2.2
Other land (000 ha) (1992-94)	3,992,533
Percentage change since (1992-94)	1.0

*Domesticated land is the sum of cropland and permanent pasture.

Source: Compiled from the World Resource Institute's Guide, 1996-97.

Among all kinds of soil degradation, water erosion is the most dominant accounting for an estimated 56 per cent of world total land degradation. Next in sensitivity is the wind erosion (28 per cent), which is prevalent in arid regions. Table 4.2 presents the status of human-induced-degraded land and degradation due to water erosion in the 1980s since this work focuses primarily at the soil erosion due to water factor.

Table 4.2: Estimates of the Land Area Affected by All Types of Human Induced Soil Degradation, 1980s

Region	*Degraded Land*		*Degree of Degradation (10^2 ha)*			
	Area (10^6 ha)	Per cent of total	Light	Moderate	Strong	Extreme
Europe	1,964.4	17	749.0	910.5	295.7	9.3
Africa	494.2	22	173.6	191.8	123.6	5.2
North America	158.1	8	18.9	112.5	26.7	0.0
South America	243.4	14	104.8	113.5	25.0	0.0
Asia	748.0	20	294.5	344.3	107.7	0.5
Ocean	102.9	13	96.6	3.9	1.9	0.4

Source: Environmental Data Report 1993-94, UNEP based on GLASOD Studies.

From the aboue table it is evident that Africa, Asia and Europe experience the greatest percentage of degradation. In absolute term, Asia faces the problem of soil degradation in around 748,000 hectare of land. In North America, South America, Oceania and Asia, extremely degrades soil are 0.0, 0.0, 0.4 and 0.5 per cent respectively, but in case of Africa and Europe it is 5.2 and 3.1 per cent respectively. This supports the argument that in case of Asia, North America, South America and Oceania the problems may be within control if efforts are made properly but it needs a massive effort for Europe and Africa.

Another global view of soil degradation is for agricultural land only; Table 4.3 gives the detail.

Table 4.3: Agricultural Land Degraded by Region (1945 to 1990)

Continent	*Share of Degraded Agricultural Land (per cent)*
Australia	16
Europe	25
North America	26
Asia	38
South America	45
Africa	65
Central America	74

Note: Area lightly degraded has been ignored here.

Source: Gary Gardener, 1997.

Table 4.4 gives the detail of degraded land due to water erosion. From the Table 4.4 it is evident that soil degradation due to water erosion is the single most dominant factor amongst all. This is true for all regions as well as the world. It is maximum for Oceania (81 per cent). It is 59 per cent and 67 per cent for Asia and North America respectively. Topsoil loss causes the greater degradation for all types of severity, *i.e.* light, moderate, strong and extreme. However, it is only in Africa that one finds extremely degraded land due to topsoil loss.

In 1991, the UN study estimated that 552 million hectares (equal to 38 per cent of today's global cultivated area) had been damaged to varied extent by agricultural mismanagement since the Second World War. As mentioned earlier, in extreme cases the farmers have abandoned the agricultural land. This type of land accounts for around 15 per cent of the world's damaged cropland (Gardener, 1997).

Table 4.4: Estimates of the Land Area Affected by Human Induced Soil Degradation due to Water Erosion, 1980s

Region/ Degradation type	*Degraded Land*		*Degree of Degradation (10^2 ha)*			
	Area (10^2 ha)	*Per cent of All Degraded Land*	*Light*	*Moderate*	*Strong*	*Extreme*
World	1093.7	56	343.2	526.7	217.2	6.6
Topsoil loss (TS)	920.3	32	301.2	454.2	161.2	3.8
(TD)	173.3	9	42.0	72.2	56.0	2.8
Africa	227.4	46	57.5	67.4	98.3	4.2
TS	204.9	41	53.9	60.5	86.6	3.8
Terrain Deformation (TD)	22.5	5	3.6	6.9	11.7	0.4
N. America	106.1	67	14.5	68.2	23.4	0.0
TS	80.9	51	14.2	60.1	6.0	0.0
TD	25.2	16	0.2	8.1	16.9	0.0
S. America	123.2	51	45.9	65.1	12.1	0.0
TS	95.1	39	34.9	51.9	8.3	0.0
TD	28.1	12	11.0	13.2	3.8	0.0
Asia	440.6	59	124.5	241.7	73.4	0.0
TS	365.2	49	99.8	215.0	50.5	0.0
TD	74.4	0	24.7	26.7	22.9	0.0
Europe	114.5	52	21.4	81.0	9.8	2.4
TS	92.8	42	18.9	64.7	9.2	0.0
TD	21.8	10	2.5	16.3	0.6	2.4
Oceania	82.8	81	79.4	3.2	0.2	0.0
TS	21.7	79	79.4	2.2	0.1	0.0
TD	1.1	1	0.0	1.1	0.1	0.0

TS—Top Soil; Loss TD—Terrain Deforestation.

Source: Environmental Data Report 1993-94, UNEP, based on GLASOD Studies.

Erosion of soil is the most prominent component of soil degradation (84 per cent). A typical hectare accumulates about one ton (a few multimeters) of new soil each year. On the contrary, reports of annual losses exceeding 100 tons per hectare are common for individual plots on sloping terrain in many developing countries. Some 160 million hectares of hillside land (more than 11 per cent of the world's cropland) were identified in 1989 as 'severely eroded' in areas like highlands of Ethiopia, the uplands of the Andes and the Himalayas, and the Central highlands of Central America.

One can estimate the soil sediment dumped into the sea as a good measure of soil erosion. In one of the pioneering assessment of soil erosion in 1968, the amount of river-borne soil sediment carried into the ocean had increased from 9 billion tons per year before the introduction of agriculture, grazing, and other activities to 24 billion tons per year (Brown & Wolf, 1984). Detailed information about the soil sediment carried by river for different countries are either obscure or subjective. Still a broad view is possible for world's Major River sediment. Following table gives the detail.

Table 4.5: Sediment Load of Selected Major Rivers in 1980s

River	*Countries*	*Annual Sediment Load (million metric tons)*
Yellow	China	1,600
Ganges	India	1,455
Amazon	Several	363
Mississippi	US	300
Irrawaddy	Burma	299
Kosi	India	172
Mekong	Several	170
Nile	Several	111

Source: El-Swaify & Dangler, 1982.

Another estimate (Walling, 1987) for soil sediment is of about 20.3 billion tons per year, which includes 15.3 billion tons of suspended sediment, 4 billion of dissolved material and 1 billion of coarser bed load. Since this accounts only for material reaching the sea and excludes for instance, sediment trapped behind the dam—it underestimates the total amount of soil lost from the land. Table 4.6 gives the detail.

The Ganges and Brahmaputra rivers jointly transport almost 3 billion tons of sediment to the Bay of Bengal each year, which is far greater than any other river system in the world. It is true that without properly quantified data, assessment of soil degradation and related issues, *e.g.* flood estimate etc. are not possible yet the fact remains that even today it is very difficult to get objective assessment of top soil loss in time-series experience.

Table 4.6: Annual Sediment Load Transported to the Sea by Major Rivers, Early Eighties

River Basin	*Country/Region*	*Drainage Area (000 km²)*	*Annual Suspended Load (Mill. Tons)*
Ganges-Brahmputra	South Asia	1,480	3,000
Yellow River	China	770	1,080
Amazon	South America	6,150	900
Chang Jiang	China	1,940	478
Irrawaddy	Myanmar	430	265
Magdalena	Columbia	240	220
Mississippi	US	3,270	210
Orinoco	Venezuela	990	210

Source: Sandra Postel, 1989.

The United States is among the few countries for which the available soil loss data commands reliability. Through the soil conservation service of 1977, the US has prepared a comprehensive inventory of soil. A second inventory, conducted in 1982, expanded the survey to nearly one million sample points nationwide — the most comprehensive look at soil resources ever completed by any country (USDA, 1982). This survey of 1982 estimated that 44 per cent of US cropland was loosing topsoil in excess of its soil loss tolerance level. The loss of soil at this excessive rate from the US cropland base of 421 million acres totalled 1.7 billion tons. With over 70 per cent coming from less than one-quarter of the cropland (USDA, 1982).

The former Soviet Union, having the largest cropland area, has also experienced the problem of soil loss. Although detailed estimates are not available but whatever indications one gets from various sources like research reports, public documents etc., the situation of soil erosion is alarming in Russia (formerly USSR) Two Soviet Scientists, Poletayev and Yashukova (1978) reported that two-thirds of the ploughed lands in the Soviet Union have been subject to the influence of various forms of erosion. The data and observations suggest that it is not unreasonable to assume that the Soviet soils are eroding at least as rapidly as those are in the United States. If 44 per cent of the land is affected by erosion at the same rate as in the US, (which may be a conservative assumption), the excessive loss of topsoil from Soviet croplands is over 2.5 billion tons per year (Brown & Wolf, 1994). China also faces the problem of soil erosion but again reliable estimates are not available. Brown and Wolf (1984) assume that China's problem is as great as that of India for which relatively better estimates of erosion are available. They write that

> "A comparison of the sediment load of the Yellow River in China with the Ganges in India indicates the relative magnitude of soil loss through erosion faced by these two population giants. The Ganges with a drainage basin of 1.1 billion square km. carries an annual sediment load of 1.46 billion tons of soil, while the Yellow River which has a drainage basin of 668,000 sq.km, carries 1.6 billion tons of soil to the ocean each year. These numbers suggest that the rate of soil loss in China is substantially greater than in India. For the purpose of constructing a rough estimate, it can be assumed that the erosion rate on China's cropland exceeds India's by 30 per cent. Given China's smaller cropland area, this means that China's excessive loss of soil from its croplands totals 4.3 billion tons per year" (Brown & Wolf, 1984).

In other underdeveloped countries, the situation is not different and soil erosion is eroding the very resource base of agriculture. Table 4.7 lists a few observations by the experts about the problem of soil erosion in these countries.

These observations, based on various sources at different point of time are enough to conclude that the problem of soil erosion is global, cutting across the political boundary and transcending into all types of economic system. They also indicate that the severity of the problem has accentuated due to the myopic vision of the planners who ignored this serious issue.

4.2 Soil Erosion in India

Before the establishment of a separate Ministry of Environment and Forest, Ministry of Agriculture used to furnish all relevant information regarding erosion and degradation of soil in India. In 1950s, Ministry of Agriculture reported that 145 million hectares of land were in need of soil and water conservation measures (MOA, 1968).

Table 4.7: Observations of Soil Erosion and Land Degradation in the Third World

Country	*Observation*	*Source*
Nepal	"Local inhabitants… all concern that the problem is more severe now than a generation ago"	Mountain Research and Development, (Boulder, Co.) 1982
Mali	"On the landsite maps, there is now—and there was not in 1976—a bright ring of soil around villages. Those areas are now 90 per cent devoid of vegetation, the topsoils is gone, and the surface is disrupted and cracked"	P.A. Jacobbenger, Smithsonian Institute, 1986.
Iran	"The area of abandoned cultivated land has doubled in recent years"	Harold Dragne Texas Tech. Univ. 1971.
Indonesia (Java)	"Soil erosion is creating an ecological emergency in Java, a result of overpopulation, which has led to deforestation and misuse of hillside areas by land hungry farmers. Erosion is laying waste to land at an alarming rate, much faster than present reclamation programmes can restore it"	US Embassy Jakarta, 1976.
Thailand	"The pace of deforestation has been accelerating since the early 1900s but it has moved into a higher gear since the 1960s (Between 1961, 1986) Thailand lost about 45 per cent of its forest"	D. Phantunvanit and K.S. Sathirathai, Thailand Development Research Board 1988.
Brazil	"Every year, rains slash deeper into the bare soil, dumping rivers to overflow into the city's streets. Now 'people' flee the drought in the North-east only to die in floods in Rio"	Helio-Pesita, Washington Post, Feb. 1988.
Mauritaia	"There are only 43 sand storms in the whole country between 1960 and 1970 the number increased tenfold in the following decade, and in 1983 alone a record 240 sandstorm darkened the nation's skies	Sidy Gaye Ambio, 1987.
Tunisia	"Rangelands have been overgrazed with three heads of cattle where only one could thrive… Two-thirds of the land area is being eaten away by desertification	UNEP, 1977.
Peru	"Erosion is estimated to affect between 50 and 60 per cent of the surface of the whole country	Mountain Research and Development (Boulder, Co. 1982).
Ethiopia	"There is an environmental nightmare unfolding before our eyes… over 1 billion tons of topsoil flow from Ethiopia highlands each year"	US AID Mission Addis Ababa 1978.
South Africa	"The province of Natal incorporating Kwazulu is losing 200 million tons of topsoil annually"	John Hanks, Institute of Natural Resources, Natal 1980.
Bolivia	"Recent aerial photographs have shown rapid extension of desert like conditions caused by wind erosion"	Helene Riviere d'Arc. Institute des Hautes, Etudes d' Amorque Latin, Paris, 1980.

Source: Compiled from the Worldwatch Institute, 1984 & 1989.

Bali (1975) estimated that 175 million hectares of land degraded (68 mha critically and 106 mha severely). The latest information furnished by the Ministry of Agriculture, Government of India is for year 1985 summarised in the following table.

Table 4.8: Areas Subject to Different Forms of Soil Erosion in India

Types of Erosion	*Area (Million hectares)*
Water	113.3
Wind	38.7
Saline alkali soils	8.0
Waterlogging soil	6.0
Ravines and guilies	4.0
Shifting cultivation	4.3
Riverine and forests	2.7
Total	177.0

Note: Around 64 per cent soil erosion is due to water borne erosion alone.

Source: Ministry of Agriculture, Government of India, 1985

Singh *et al.* (1983) in a separate study have found that about 5334 million tons of soil (16.4 mt/ha) is eroded every year. The country's rivers carry an approximate quantity of 2052 Mt of soil (6.02 t/ha). Of this about 29 per cent is lost to the sea, 10 per cent is deposited in reservoirs and 61 per cent is transported from one place to another.

A comprehensive look of the soil erosion locating major areas has been given in Table 4.9.

In more recent studies Singh, Rambabu, Narain, Bhushan and Abrol (1992) have estimated the soil erosion for India using universal soil loss equation (USLE) and soil erosion maps at 1:6,000,000 scale. Their findings have been summarised in Table 4.9.

The maximum area of degradation falls under moderate scale and very severely eroded land is the minimum. Singh *et al.*, (1992) clearly indicate that erosion rate in the black soil regions of the country, occupying 64 million hectare area in Karnataka, Andhra Pradesh, Madhya Pradesh and Maharashtra states was 20 mg/ha. Red soil of Chhotanagpur plateau also recorded a soil loss value of 10 to 15 mg/h/yr.

Table 4.9: Estimated Soil Erosion (in tones) per annum by Water in Different Land Resources Regions of India

Land Resource Region	*Area (km²)*	*Soil Eroded (t/km²)*	*Soil Eroded in the Region (Million tons)*
A. North Himalayan Snow clad	116000 (3.56)	—	—
B. North Himalayan Alpine Grass and Meadows	98250 (3.02)	15.93	1.6
C. North Himalayan Forest	131751 (4.08)	287.0	28.4
D. Punjab-Harlan Alluvial Plain	101250 (3.1)	330.0	33.4
E. Upper Gangetic Alluvial plain	200000 (6.15)	141.0	242.7
F. Lower Gangetic Alluvial plain	145500 (4.41)	940.0	132.3
G. NE Himalaya Alpine grass and meadow	16000 (0.50)	50.0	0.6
H. NE Himalayan Forest	161000 (4.97)	287.0	23.2
I. Assam Valley	88500 (2.73)	2815.0	249.1
J. Rajasthan Desert	191000 (5.87)	NA	NA
K. Rann of Kutch	46500 (1.44)	NA	NA
L. Gujarat Alluvial	62750 (1.94)	480.0	28.2
M. Mixed yellow, Red and black Soil	115750 (3.08)	300.0	53.4
N. Black Soil	673500 (20.73)	6448.0	3375.5
O. Eastern Red Soil	573500 (17.65)	346.0	141.2
P. Ganges Delta	25250	—	
Q. Western Coastal	6100 (1.88)	3930.0	234.8
R. Red Soil	347750 (10.72)	359.0	95.0
S. Eastern coastal	93500 (2.9)	3930.0	367.5
Roadside erosion	10000	1969.0	19.0
			5364.4

Source: Khosoo & Tejwani, 1993.

Table 4.10: Area under Different Classes of Soil Erosion by Water in India

Soil Erosion Rate	*Area (km²)*	*Soil Erosion Class*
Range (Mg/ha/year)		
0-5	801,350	Slight
5-10	1,405,640	Moderate
0-20	805,030	High
20-40	160,050	Very high
40-80	83,300	Severe
More than 80	31,895	Very severe

Source: Singh *et al.*, 1992

The northwestern hill or Jammu and Kashmir, Himachal Pradesh and Uttar Pradesh and north-eastern hills of Bengal and the north-eastern states contribute more than 20 mg/h/yr. Fronthills of the Himalayas and the Doon Valley produce 20 mg/h/yr. of soil loss.

4.3 Global Assessment of Soil Degradation (GLASOD)

International Soils Research and Information Centre (ISRIC) under the aegis of UNEP has also conducted the study for soil degradation for India under the project Global Assessment of Soil Degradation (GLASOD). GLASOD has used 26 units of maps and 33 records for assessment of all types of soil degradation for India. GLASOD studies also take into account of the studies done by the Government of India and individual experts. However, due to the difference in the severity measurement, estimates may vary from the GLASOD estimates by a factor of two and sometimes even more (FAO, 1994).

Following tables provides the detail about all kinds of soil degradation for India.

Maximum damage to Indian agriculture land is by water erosion. Wind erosion is dominant in dry regions and water erosion in humid ones.

From the GLASOD studies and other individual studies, it is obvious that they differ significantly, sometimes the variation is even greater than 100 per cent. On *a priori* basis one can't prefer a study to another. Even the latest and most sophisticated study done by GLASOD is not a sacrosanct, the reasons are many but most important is that the time-series data on land degradation is scarce and unreliable. Even today, data on land degradation is controversial. And in 1950s information on soil conditions was simply absent for most of the nations. The GLASOD Study therefore

Table 4.11: GLASOD Estimates of Land Degradation for India

Type		*Degree of Degradation*				*Total as a Percentage of Agricultural Land*
		Light	*Moderate*	*Strong*	*Total*	
Water	*Erosion*					
Dry	Region	1.2	0.0	1.7	2.9	
Humid	Region	1.8	17.2	10.9	29.9	
Total		3.0	17.2	12.6	32.8	18 per cent
Wind	*Erosion*					
Dry	Region	0.0	1.8	9.0	10.8	
Humid	Region	0.0	0.0	0.0	0.0	
Total		0.0	1.8	9.0	10.8	6 per cent
Soil Fertility	*Decline*					
Dry	Region	2.2	0.0	0.0	2.2	
Humid	Region	24.2	0.0	3.2	27.2	
Total		26.2	0.0	3.2	29.2	16 per cent
Lowering of the water table						
Dry	Region	0.1	0.1	0.0	0.2	
Humid	Region	0.0	0.0	0.0	0.0	
Total		0.1	0.1	0.0	0.2	<1 per cent
Water	*Logging*					
Dry	Region	0.0	3.1	0.0	3.1	
Humid	Region	0.0	0.0	0.0	0.0	
Total		0.0	3.1	0.0	3.1	2 per cent
Salinization						
Dry	Region	0.0	3.5	3.5	7.0	
Humid	Region	0.0	0.0	0.0	0.3	
Total		0.0	3.5	3.5	7.3	4 per cent

Source: FAO, 1994.

relies heavily upon the judgement of 250 soil scientists, which becomes a subjective issue (Dyson, 1998).

5

Socio-Economic Causes of Soil Erosion

5.1 Natural Causes of Soil Erosion

Erosion of soil is caused by a myriad of factors ranging from natural to cultural one. Interplay of diverse drivers arising out of entirely different contexts and situations, evolve a resultant force, which ultimately create the phenomena of soil erosion. Natural factors acting as catalysts in the process of soil degradation can be classified under different kinds of degradation. Natural factors responsible for water erosion are:

5.1.1 High Intensity Monsoon Rains

Soil loss takes place with rainfall intensity. Raindrops detach the soil of surface. This applies particularly to erosion by overland flow and rills for which intensity is generally considered to be the most important rain characteristic.

5.1.2 Steep Slopes of Mountain and Hill Lands

Erosion of topsoil would normally be expected to increase with increase in slope steepness and slope length as a result of respective increase in velocity and volume of surface runoff. The relationship between the erosion and slope can be expressed as

$$E \propto \tan \theta L$$

where, E = Soil loss, θ = Slope angle, L = Slope length

Natural hazards of soil degradation are of minor concern for relevant time horizon of human economic activity. They do not change significantly unless they are intervened with all intensity consciously or unconsciously by the man-made factors. Characteristics of rain drops (velocity, drop size etc.) and steepness or the slopes does not change rapidly and hence can safely be presumed to be constant for a given 40 to 50 years time framework for economic and other social policies.

5.2 Direct Causes of Soil Degradation

Soil erosion takes place mostly due to excess pressure of homosapiens on the natural vegetation in order to cater to growing demands for food, fuel, fodder and fibber. These factors can be called as direct causes of soil degradation. They are as follows:

5.2.1 Deforestation

Cutting down the trees for agricultural purpose exposes the land to hostile natural conditions (*e.g.* rain, wind etc.) and causes soil loss. Deforestation mainly causes water erosion and becomes really severe if the cleared land lies along the steep slope. In last 300 years a sizeable amount of land has been cleared for agricultural purposes causing pressure on forest area of the world.

In an estimate done by FAO, it is observed that each year 11.3 million ha of tropical forests are lost through the combined action of land clearing for crop production, fuel wood gathering and cattle ranching. However, recent satellite data for individual country put this FAO figure as the most conservative one.

This alarming rate of deforestation and overcutting of vegetation lead to the erosion of topsoil at the global level.

5.2.2 Shifting Cultivation Without Adequate Fallow Periods

Shifting cultivation has been one of the most dominant ways of agriculture in the past with adequate fallow period allowing the soil to regain its fertility but explosion in population and subsequent enhancement of man-land ratio reduced the fallow period. Eckholm (1976) writes:

> "Shifting cultivation is often seen as a stigma of

> backwardness, but where the ratio of land to people is high enough, it is an ecologically sound system and one well adapted to tropical forest conditions. Few of the men and women who practice it have been educated in primary schools, let alone universities. Yet they usually carry in their head an extraordinary found of scientifically sound knowledge about plant species and soil qualities. Despite the ecological soundness, however, shifting cultivation has serious disadvantages. Production may be fairly secure, but the system makes no provision for the accumulation of investments which is the key to higher productivity and wealth. Most societies try to mould the environment in directions that serve human ends but shifting cultivation implies the almost complete perturbations with human practices to fit the environment as it is. But most importantly, in many regions the system can support only an extremely sparse population. Though, conditions vary widely, often at least fifteen hectares of land must be available for each dependent person if shifting cultivation is to be sustained over time. Once a threshold point is reached at which farmers returns to a plot before its fertility is fully restored, a dangerous cycle of degradation begins, unless skilful and painstaking measures to recycle nutrients and organic matter to the soil are scrupulously carried out."
>
> (Eckholm, 1976, p.139)

Even when the normal agricultural practice is followed proper fallow time allows farmer to work on land both in semi-arid regions and in the tropics where nutrients are scarce. In recent years, rising demand for food has reduced the fallow land considerably. In the US, as the world wheat prices climbed sharply during the mid-70s, the summer fallow land dropped from 17 mill. Ha in 1969 to 13 mill. ha in 1974 (UDSA, 1981, 1982). In the former USSR, during the late sixties and early seventies, the Soviets were consistently fallowing 17.8 million hectares each year in the dry land regions. After the huge gap in the supply of foodgrains in 1972 the followed area reduced by one-third (USDA, 1983). In most of the third world countries, the fallow period has reduced owing to the greater need for food and other primary produce.

In India, shifting cultivation is found in the hilly areas of north-east India, where it is a cause of water-induced erosion and soil fertility decline.

5.2.3 Overgrazing

Overgrazing means the grazing of natural pastures above its livestock carrying capacity. It causes the decline in the quantity and quality of vegetation cover. Excess of the livestock population exposes soil to wind and rain. These animals also trample the surface and either dislodges soil particles or compact the surface reducing infiltration. In a study done for Swaziland (Gambia), it has been found that intensity of grazing significantly reduces the soil life. Table 5.1 shows the complete result. The rates and pattern of damage depends upon the variety of livestock.

Table 5.1: Grazing Intensity and Life Expectancy of Soil, Swaziland (Now in Gambia)

Estimated Grazing Intensity (Ha/animal unit)	*Area Affected (ha)*	*Life Expectancy of Soil in Year*
0.40-0.75	6,790	less than 5
0.75-1.00	55,220	5-10
1.00-1.25	65,200	10-15
1.25-1.50	91,600	15-25
1.50-2.00	79,355	25-50
2.00-2.50	2,56,190	50-100
Over 2.50	6,90,165	100

Source: Roder in Blaikie (1985).

5.2.4 Extension of Agriculture onto Lands of Lesser Suitability

Growing demand for food pushes the arable land into less suitable areas known as 'fragile' or 'marginal' land. In a typical Ricardian pattern one can discern where more fertile land are domesticated first and then the second best. Such marginal lands could be:

(a) Steeply sloping land
(b) Areas of shallow or sandy soils
(c) Cultivation of semi-arid land, and grazing the drier semi-arid areas, marginal to deserts.

Agriculture on such land may increase the output but causes serious damage to the soil. In last hundred years or so, mountainous regions of Japan, China, Nepal, India and Indonesia have been occupied by the poor people (mostly migrants). Growing competition among the farmers for cost efficiency does not allow proper terracing causing frequent landslides. A study for Nigeria has shown that agriculture on a steeper slope cause greater loss of soil for a given variety of crops. Cassava planted on land of 1 per cent slope costs an average of 3 tons of soil per year against 87 tons per year on a 5 per cent slope land—a rate at which topsoil layer of six inches would disappear entirely within a generation. Further, cassava planted on a 15 per cent slope would lead to an annual erosion of 221 tons per hectare—a rate enough to erode all the top soil within a decade (Aina, 1982).

It is noteworthy that such fragile lands need greater effort and vision for conservation practices but unfortunately as they are owned by poor farmers who are extremely deprived and hence lack any motivation for conservation endeavour.

5.2.5 Improper Crop Rotations

Demographic pressure demands continuous growth in food production, which are generally realized by intensive agriculture. Such agriculture is cereal producing mainly rice and wheat. Not much empirical evidences are available but majority of the soil scientists agree that monoculture type of agriculture cause more harm to topsoil. Research has shown that in Missouri (US) during thirties, an increase in soil erosion from 2.7 tons per acre annually, when land was in a corn-wheat-clover rotation, to 19.7 tons per acre when the same land planted continuously to corn. Table 5.2 shows the details.

Table 5.2: Cropping Systems and Soil Erosion

Cropping System	*Average Annual Soil Loss (T/acres)*
Corn, wheat and clover rotation	2.7
Continuous wheat	10.1
Continuous corn	19.7

Source: Miller, 1980.

Generally, the loss in soil and subsequent fertility loss is compensated by using higher doses of chemical fertilizers.

5.3 Socio-Economic Factors—Underlying Causes of Soil Erosion

Natural causes as well as direct causes of soil erosion emanate from the social and economic fabric of the system in which human beings operate. Deforestation arising out of rapidly increasing demand for food, fuel and other fibbers in the major part of the world is the result of a complex interaction of poverty, population, political system, pattern of ownership of land and forest, class interest, historical experiences, ethical and religious beliefs of the people.

5.3.1 Ever Increasing Population—Land Ratio and Unsustainable Bovine Population

In last 300 to 400 years, the global population has exploded at an alarming rate, especially after the Second World War. In last 50 years the population of Europe and America has got stabilized but that of Africa, Asia and Central America continue to grow at an unparalleled rate. Global population was supposed to be around 680 million in 1700 AD and will be reaching to around 8000 million in 2020 AD Table 5.3 shows this trend.

Table 5.3: Global Population (millions) 1700-1985

Year	*Population*
1700	679
1750	769
1800	957
1850	1260
1900	1650
1950	2515
1985	4853
2020	8061

Sources: 1. Turner II, Clark *et al.*, 1990 and 2. Miller, 1980.

Areawise distribution shows that the maximum explosion of population has taken place in Europe and America (North) during 1700-1900 and in Africa, Asia and Latin America during 1900-1985 (Table 5.4).

Table 5.4: Areawise Distribution of Population in millions (1700-2000)

Region	*1700-1750*	*1750-1800*	*1800-1850*	*1850-1900*	*1900-1950*	*1950-1985*	*1985-2020*
World	90(.25)	188 (.44)	303(.55)	390(.54)	865(.84)	2338(1.88)	3208(1.45)
Asia	63(.3)	132(.5)	171(.5)	124(.3)	450(.8)	1459(2.1)	1846(1.4)
China	—	123(1.0)	107(.6)	6(.0)	119(.5)	505(1.9)	400(.9)
Japan	—	0(.0)	1(.1)	13(.7)	40(1.3	37(1.0)	9(.2)
Rest of Asia	—	9(.1)	63(.4)	105(.5)	291(.5)	9.7(2.3)	1437(1.8)
Africa	1(.0)	1(.0)	4(.1)	22(.4)	91(1)	333(2.6)	884(2.7)
Europe [a]	17(.3)	36(.6)	63(.7)	86(.7)	99(.6)	99(.6)	22(.1)
USSR	5(.3)	14(.7)	30(1.0)	48(1.0)	53(.7)	97(1.2)	66(.6)
North America [b]	1(.8)	2(1.0)	20(3.2)	5(2.6)	76(1.2)	99(1.3)	62(.6)
Latin America [c]	5(.8)	4(.5)	15(1.2)	41(1.6)	90(1.6)	239(2.6)	315(1.6)
Ocean	0(-)	-1(-)	0(-)	4(-)	7(1.6)	12(1.9)	12(1.2)

In bracket show the average annual rates of population increase in per cent.
a—excluding the USSR, b—Canada and the US, c—South and Central America, Mexico and the Caribbean — Not available
Source: Compiled from Turner II *et al.*, 1990.

From the table it can easily be discerned that Africa, Latin America and Asia are regions of greater increase in the population growth rate. In Asia, rest of Asia is dominated by Indian subcontinent where growth of population is continuously increasing till 1985. During 1985-2020 the projection shows some decline for some regions but upto 1985 they have increased at a very high rate.

Growing population needs area for agriculture, recreation and shelter and has caused clearing of forests and woodlands resulting into decreasing per capita forest area availability. Table 5.5 shows the declining ratio of forest and woodland area to population at global level.

Table 5.5: Global Population and Land Use Pattern 1700-1980

Year	*Population (million)*	*Forest and woodland (mill. Ha)*	*B/A (ha)*
	A	B	
1700	679	6215	9.15
1850	1260	6965	5.53
1950	2515	5389	2.14
1980	4325	5053	1.17

Source: Derived from the tables in Turner II & Clark 1990.

Continuously declining per capita availability of forest and woodland is both a cause and impact of global soil erosion. Growing pressure on land and requirement for food has compelled humans to go for cultivation in steep slopelands and semi-arid and arid areas of very high erodivity.

In India the population has increased from 238.4 mill. in 1901 to 846.3 million in 1991, but the cropland has increased only, marginally from 131.9 mill. ha in 1950-51 to 185.5 mill ha. in 1991, reducing the per capita availability of cropped area (Table 5.6).

Table 5.6a: Global Per Capita Availability of Cropland

Year	*Population*	*Gross Cropland Area*	*B/A (ha)*
1700	679	625	0.390
1850	1260	537	0.430
1950	2515	1170	0.465
1980	4325	1501	0.347

Source: Derived from the tables in Turner II *et al.* 1990.

Table 5.6b: Per Capita Availability of Cropland in India

Year	*Population (million)* A	*Forest Area (Miil. ha)* B	*Cropped Area (Miil. ha)* C	*B/A (Ha)*	*C/A (Ha)*
1951	361.10	40.5	131.90	0.112	0.365
1961	439.20	54.0	152.78	0.13	0.348
1971	548.20	63.9	165.80	0.116	0.302
1981	685.20	67.4	177.40	0.098	0.259
1991	846.30	68.3	185.56	0.080	0.219

Source: Indian Agriculture in Brief, 24th edition, 1992, Fertilizer Statistics, 1993-94, Report of the National Commission on Agriculture, Forestry Statistics, 1995.

Per capita availability of forest and cropped area is declining in India and presuming over zero deforestation rate in future, this trend will continue owing to the population growth. Greater density of population plays havoc in the hilly regions of the country. Generally, it is said that slopelands are sparsely populated, which is true if total area of land is taken into account. The density of population in the entire Indian Himalayan is 72 persons per km^2 as compared to the all

India figure of 211 persons per sq. km. If the population is counted on the basis of area in cultivation it comes to be 1432 persons per Km^2 in the entire Himalayas as compared to 483 persons per km^2 for the whole of India (Khosoo and Tejwani, 1993).

Indian bovine population has also been increasing continuously. India has the largest bovine population in the world. Notable among them are the number of sheep and goats which cause maximum damage to the soil and lead to the process of desertification.

Table 5.7: Livestock Population: Trends (millions)

Item	*1951*	*1961*	*1971*	*1981*	*1991*	*1993*
Buffaloes	40.83	51.21	57.43	63.87	78.96	78.55
Ships	36.83	40.02	39.99	41.33	44.21	44.61
Goats	45.15	60.86	67.52	76.90	116.75	117.55
Horses & Ponies	1.41	1.33	0.94	0.95	0.96	0.98
Mules	0.06	0.069	0.07	0.14	0.14	0.14
Asses	1.25	1.02	0.99	1.45	1.50	1.55
Camel	0.64	0.90	1.11	1.05	1.49	1.51

Source: FAO Yearbook 1969, 1976, 1981, 1985, 1987, 1992 & 1993.

The population of goats and sheep's has grown significantly by around 25 per cent and 150 per cent respectively between 1951 and 1993. This growth exceeds the pastoral's carrying capacity and cause overgrazing which leads to a considerable soil erosion.

5.3.2 Fragmented Land and Declining Farm Size

Ever increasing population has reduced the farm size. Small land holdings lead to severe economic pressures on farmers, to obtain sufficient food and income to meet immediate needs. This creates problem in the way of efficient land management. In India, mismanagement of land has largely raised out of growing small size land holdings. Table 5.8 shows the increasing fragmentation of land size.

Average size of holding is continuously declining from 2.3 million in 1970-81 to 1.7 million in 1985-86. Land holding of sizes less than one hectare increased from 36.20 mill. in 1970-71 to 56.15 million in 1985-86.

Table 5.8: Size-Distribution of Operational Holdings in India

Size Group	*Number of Holdings (million)*			
	1970-71	1976-77	1980-81	1985-86
Less than 1 ha	36.20 (51.0)	44.52 (54.5)	50.12 (56.4)	56.15 (57.8)
1-2 ha	13.43 (18.9)	14.73 (18.1)	16.07 (18.1)	17.92 (18.4)
2-4 ha	10.68 (14.9)	11.67 (14.3)	12.45 (14.0)	13.25 (13.6)
4-10 ha	7.93 (11.2)	8.21 (10.0)	8.07 (9.1)	7.92 (8.2)
10 and above	2.77 (3.2)	2.44 (3.0)	2.17 (2.4)	1.92 (2.0)
Average size	2.3	2.0	1.8	1.7

Note: Figure in brackets shows percentage to the total cultivable land.

Source: Indian Agriculture in Brief, 1992.

This trend is especially damaging the topsoil along the slopes. In Uttar Pradesh Himalayas, 70 per cent of the farm holdings are less than 1 ha and only 13 per cent are more than 2 ha. In small watersheds in Almora district, 75-83 per cent of holdings are less than 0.5 ha (Khoosoo & Tejwani, 1993). Additional feature of this small size of land holding is that they are scattered in different locations. The number of locations for this same district of Almora could be up to 13. This phenomena transcends into social system from the economic one but ultimately causes great loss to the soil.

5.3.3 Land Tenure System

Generally, farmers are reluctant to investment in land conservation programmes, which check the soil erosion; if the land they are using does not belong to them. This situation may arise in case of absentee tenancy and open access resources. In a study for Monroe Country, Missouri (USA), it has been found that there are significantly less erosion control and conservation practices followed on rented cropland than a cropland formed by owner operations (Ervin, 1982). In the US about 40 per cent of total cropland is rented. Colonial political system in large part of Asia, Africa and Central America has perpetuated this problem and caused the degradation of soil, which is unparalleled in the history. Landowners are from the cities and the cultivators are the locals. These local people who are invariably extremely poor do not see beyond their short time-horizon and believe in the immediate return only. Even after the political independence of these countries, the situation has not improved much. India possesses possibly one of the largest number of landless labour in the world in spite of Government's land distribution programme. During 1985-86, there were around 53,899 thousand of marginal and 17,922 thousand small holdings operating for 47,717,000 hectares of land—quite a significant fraction of total land area. A significant percentage of that is a leased one (Agriculture Census, 1985-86).

5.3.4 Cumulative Impact of Poverty, Marginalisation of People and Lopsided Development Pattern

Most of the developing and underdeveloped economies face the chronic problem of poverty. Their farmers and land users place highest priority to the basic necessities of life. Their number is large and they are marginalised and hapless in the distribution of the fruits of development in their country. Small farmers make a large and direct contribution to the land degradation in many watersheds. Southgate (1988) describes—"practicing erosive agriculture in one place accelerates the rates at which soil and gravel are scoured from streambanks and roadside downhill. Directly and indirectly, then small farmers are primary agents of land degradation in developing countries."

People living below the poverty line who are in very high proportion of total population in most of the third world countries including India's are both the result and cause of the process of soil degradation. Small marginal farmers who closely interact with the land resources and depend upon them directly for their livelihood act 'undesirably', cause the erosion and become poorer as the source of welfare becomes inferior. The interactive process of poverty and soil degradation in the underdeveloped countries has been shown in the following Fig. 5.1.

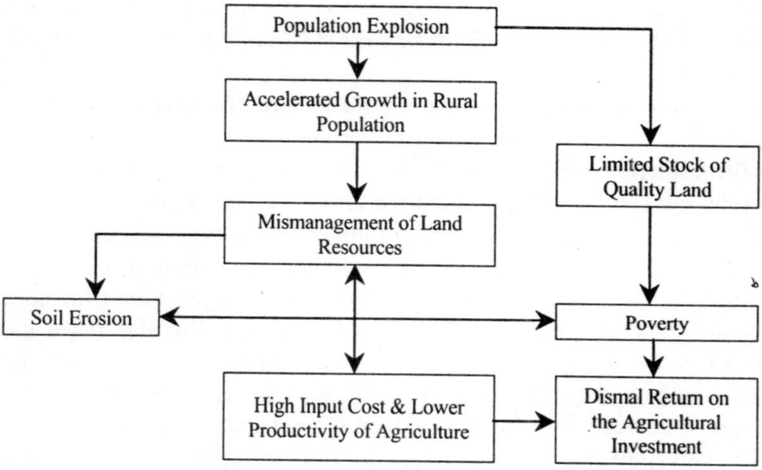

Fig. 5.1: Poverty Soil Erosion Linkages

Rapid growth in rural population makes the size of operational holding very small. Farmers in the third world countries face an alien type of production system emphasizing on mechanized and large-scale operation. They also find themselves unable to participate into modernized and Eurocentric development mode due to the lack of training and their preconceived value system. This process creates a large surplus of labour force in the urban area and finds no alternative except migrating to the countryside and cultivating fragile ecosystem land besides adding to the already existing stock of rural labour force.

Transgressing the link of poverty and soil degradation to the political economy context, Blaikie (1985) contends that agricultural colonization of hinterlands is also observed because large farmers influence the market and tenurial environment. He further states that there are signals projected through the donor agencies, research centres, and government bureaucracies in most of the developing countries that there is no alternative to develop large-scale farming system for rapid growth (Pesner and McPherson, 1985). As large farmers adopt such system, displacing more and more peasants. Blakie argues that concentrated economic interests promote the rural poor's settlement on small parcels in order to reduce wages paid to unskilled workers in agriculture and other sectors. Thus the recognition of this consequence of creating a class of small landowners in the countryside accentuates the soil erosion (De Janvry 1981). One can also find a similar case in Columbia (Taussing, 1978).

5.3.5 Distorted Macro Economic Policies and Misplaced Goal

Often, interplay of different imperfect markets (labour, capital, goods, credit etc.) in developing countries throw wrong signals in terms of distorted prices for different commodities leading to their over exploitation (under exploitation). General equilibrium of fiscal, monetary and trade policies generally do not recognise their linkages to soil and their quality. Politically motivated agricultural policies and other sectoral planning either project a wrong signal or do not acknowledge their impact at all. In these countries, preconditions for market mechanism to function smoothly are hardly found.

Government's induce imperfections in terms of taxes and subsidies, which depend upon the vested interest of the dominant class of the country concerned, further aggravate the problem. Government's interventions due to different ideological convictions distort the commodity prices, which may function against the spirit of land conservation programmes by the farmers.

In 1980s, commodity prices declined in Ecuador. This fall was caused largely by the fact that Ecuador could not insulate itself from strong trends toward labour prices in international market (Southgate & Whitaker, 1994). Subsidies in rich countries for agricultural produce and exports and increase in yield caused decline in the real grain price by more than a third during the 1980s (FAO, 1992). This induced a preference for growing crops like beans and potatoes. In addition, output of cut flowers, asparagus and other speciality crops in North America and other markets increased. These new acquired crops caused more soil erosion in the highland Ecuador.

Livestock production also underwent a significant change in Latin America, which demonstrates how the distorted price can accelerate the soil erosion. Estrada (1993) contends that the conversion of cropland to pasture, which has affected millions of hectares of land from Bolivia to Colombia, can easily be explained primarily by the fact that the prices of milk, meat and other livestock products have risen relative to the market value of crops. This is due to the enhanced demand for livestock products caused by the growth in income and public policy where basic grain prices has been controlled and dairy processors face little monitoring. Growing price for dairy products has led to the accelerated growth in cattle ranching leading to overgrazing and hence soil erosions. One can find plenty of instances of distorted fiscal and monetary policy in large number of countries in Africa, Asia and Latin America where it has caused erosion of topsoil.

5.3.6 Social Norms, Code of Conduct and Value System

Many times special social tradition and code of conducts also help in land degradation. Different types of rituals and position of women in the society etc. become inconducive to soil management practices. Overgrazing has been identified as one of the major reason for soil erosion (60t/ha/anum) in Sukumaland in Tanzania. The Tanzanian Government in 1980 recommended a 10 per cent destocking of the rangeland and compulsory tree planting. Farmers responded by hiding their cattle and neglecting the springs. Low-level aerial survey revealed more than two million cattle in just two of the most degraded districts of Sukumaland, nearly twice the official figure, arrived at by ground census. People's attitude behind this kind of non-cooperating behaviour can at best be explained in their perception and exposure to life and material well-being. Michael Stocking (1988) extends the explanation "there are few things to buy in the shops of Shinyanga, the regional town. Ox-drawn ploughs are rationed and new suit unobtainable. Cattle on the other hand, are many. They provide insurance in hard times, a walking bank in times of plenty, they are a symbol of wealth and prestige, an asset to exchange for an additional wife."

One can trace similar social stigma and their own predilection for well being in many societies, which accelerates the process of soil degradation.

6

Impact of Soil Erosion and its Economic Implications

While soil erosion may be a natural phenomenon accelerated by human-induced factors, its implications are purely economic affecting the material life and overall well being of the society. Soil erosion of all ranges, *i.e.* light, moderate, severe and extreme inflicts heavy costs on the producers and sometimes on the household sectors. In order to gauge the economic implications of erosion on the society, it would be imperative to analyse its impact in detail.

6.1 Impact of Soil Erosion

Impacts of soil erosion are generally classified into two categories:

(i) On-site (direct) Impact, and (ii) Off-site (indirect) Impact.

6.1.1 On-site (direct) Impact

On-site impact is localised and pertains to the eroded land only. Generally, it is referred to as direct impact of erosion. Under on-site impact, productivity decline is the most dominant.

Decline in Productivity of Crops: Erosion of soil reduces the nutrients of the soil along with loss in organic matters. It also reduces the water holding capacity of the soil, which is crucial for the growth of plants. These impacts impair the growth of the crops and hence productivity. However, in spite of the fact that soil erosion depletes soil productivity and plant growth, the relationship between erosion and productivity is not well defined (Landgale & Strader, 1998; Lyles, 1975; Pesek, 1980; Pimentel 1976; Stout, 1965; Crosson, 1997). For plant growth, combinations of some soil attribute like soil depth, root zone, water storage and transmission capacities, chemical/nutritional qualities and biological qualities etc. are needed. If the ideal rooting characteristics of a crop and its soil-based requirements for optimum production are known, erosion impacts on productivity can be predicted with some certainty from soil-loss data. Several predictive tools have been designed to accomplish this task but their applicability remains largely restricted to temperate areas (Larson *et al.*, 1985, Williams, 1985). Of late, a few research studies have shown that erosion impacts on the productivity of highly weathered tropical soils are more severe than for temperate soils, because the higher erosion rates in the humid tropics causes more severe damage to organic matter and other nutrients. A poor farmer of these regions cannot bring back these nutrients.

A few empirical studies showing the relationship between erosion and productivity loss have however boosted the arguments that in all regions and under all types of erosion, productivity of crop declines. National Soil Erosion-Soil Productivity Research Planning Committee of the US reported that when corn was grown continuously on a plot in Iowa from which the topsoil had been removed, yields were only 20 per cent of those on a control plot. By using most comprehensive data Leon Lyles (1975) compared 14 independent studies mostly undertaken in the corn states (Table 6.1).

Table 6.1: Effect of Topsoil Loss on Corn Yields

Location	*Yield*	*Reduction Per Inch of Topsoil Lost*	*Soil Description*
	(Bushel/acre)	(Per cent)	
East Central, Illinois	3.7	6.5	Swygert silt loam
Fowler, Indiana	4.0	4.3	Fowler, Brookstone & Parr silt loam
Clarinda, Iowa	4.0	5.1	Marshall silt loam
Greenfield, Iowa	3.1	6.3	Shelby silt loam
Shenanadoah, Iowa	6.1	5.1	Marshall silt loam
Bethany, Missouri	4.0	6.0	Shelby and Groundy silt loam
Columbus, Ohio	3.0	6.0	Celina silt loam
Woosfer, Ohio	4.8	8.0	Canfield silt loam

Source: Leon Lyles (1975).

Lyles found that soil loss reduced yields by as little as 3 bushels per acre to as much 6.1 bushels per acre. Lyles further argues that the loss of an inch of topsoil reduced corn yields at those 14 sites by an average of 6 per cent. Lyles extended his compilation exclusively for wheat and concluded that the loss of an inch of soil reduced wheat yields by an average of 6 per cent (Table 6.2).

Table 6.2: Effect of Topsoil Loss on Wheat

Location	*Yield Reduction Per Inch of Topsoil Lost (Bushel/acre)*	*Per cent*	*Soil Description*
Akron, Colorado	0.5	2.0	Weld silt loam
Geary county, Kansas	1.3	6.2	(not available)
Manhattan, Kansas	1.1	4.3	Smolan silty clay loam
Columbus, Ohio	1.3	5.3	Cropped soil
Oregon	1.0	2.2	Deep soil
Oregon	2.5	5.8	Thin soil
Palouse area, Washington	1.6	6.9	(not available)

Source: Leon Lyles (1975): Journal of Soil and Water Conservation.

In India, since 1950s studies have been conducted to estimate the impact of soil erosion on nutrient loss, water loss and overall productivity decline. It is estimated that 5.37 to 8.4 Mt of plant nutrients are lost every year due to soil erosion (MOA, 1985). Various estimates have suggested that when 2.5 Cum of topsoil was removed artificially, 14 per cent individual decrease was observed in the yield of maize-grain and stalk, when 75 Cum of topsoil was removed, the maize-grain and stalk yield were reduced by 33 and 27 per cent respectively. In various parts of the country, loss of rooting depth has also been reported—an essential feature for plant sustenance and growth. Due to soil erosion and loss of organic matter, the subsoil has lesser capacity to hold the water. As a consequence of these two processes soil cannot support good plant growth. This phenomenon is more prevalent in the hills of Himachal Pradesh and Uttar Pradesh.

Loss of organic matter for Indian subcontinent is a matter of serious concern because no chemical fertilizer can fully compensate for the loss of organic matter. In Indian subcontinent and in India due to extreme poverty, crop residues are widely used as fuel and fodder, which are not returned to the soil. In Bangladesh, the average organic matter of topsoil is said to have declined by 50 per cent (from 2 per cent to 1 per cent) over the last 20 years (Bangladesh, 1992). In Haryana soil test reports of over 15 years now show a decrease in soil carbon (Chaudhary and Aneja, 1991). Nutrient depletion is also prevalent in India (Biswas and Tewatia 1991; Tandon, 1992). Growing use of fertilizers for more or less the same level of yield also testifies the declining productivity of crops in India. In India, fertilizer use is dominated by Nitrogen. The N.P.K. (Nitrogen, Phosphate, Potassium) ratio for India is 1.00:0.33:0.17 compared with 1.00:0.52:0.40 for the world (Pradhan, 1992). This trend goes back to the green revolution period in the 1960s when fertilizers were first applied to a soil and a high response was frequently obtained from nitrogen. For both major cereal crops of wheat and sugar, to maintain a given yield higher doses of nitrogenous fertilizers have been used continuously, thus, kilograms of wheat and rice for each kg of nitrogen are declining.

Table 6.3: Kilograms of Wheat and Rice Produced per Kilogramme of Nitrogen Supplied as Fertilizers

Year	*Wheat (Million tons) (A)*	*Rice (Million tons) (B)*	*Concentration of Nitrogenous Fertilizer (Million tons) (C)*	*A/C Wheat (kg) per kg. of N. Fertilizer*	*B/C Rice (kg) per kg. of N. Fertilizer*
1960-61	11.0	36.4	0.210	52.38	173.33
1970-71	23.8	42.2	1.487	16.00	28.38
1980-81	36.3	53.6	3.678	9.87	14.57
1990-91	55.1	74.3	7.997	6.89	9.29
1996-97	69.3	74.7	10.302	6.73	7.23

Source: Compiled from the Economic Survey, Government of India, 1997-98.

At a broader level El-Swaify (1993) has compiled the data related to soil erosion in the tropics and their impact on the productivity. A few remarkable observations by El-Swaify are as follows:

(a) Erosion effects on productivity differ for different crops. Evidence so far is that crop sensitivity is in the order grain: cereals > grain legumes > root crops.

(b) Erosion effects are modified by the level of inputs (Fertilizers, tillage, supplemental irrigation etc.) provided to the growing crops.

In a study, Paul Rosenberry *et al.* estimated that as soil progresses from moderately to severely eroded category, the amount of nitrogen, phosphorous and potash needed to grow corn increases by 37 pounds per acre (Table 6.4).

Table 6.4: Effect of Topsoil Loss (Scale-wise) on Corn Yields

Change in Erosion Phase	*Nitrogen*	*Phosphate*	*Potash*
Slight to moderate	10	2	6
Moderate to severe	30	1	7

Source: Paul Rosenberry *et al.* (1980).

Use of 1 kg nitrogenous fertilizer each could produce 52.38 kg of wheat and 173.33 kg of rice respectively in 1960-61, which came down to 6.73 kg and 7.25 respectively in 1996-97. While there could be several other reasons, a major contributory factor is undoubtedly the decline in soil productivity.

(c) The replenishment of soil nutrients (as added fertilizers) by itself is not sufficient to restore original soil productivity. Physical and biological rehabilitations are also required.

(d) Highly weathered tropical soils are more vulnerable to erosion effects than temperate soils.

(e) Rooting depths in eroded tropical soils are considerably more restricted than in uneroded soils.

(f) Unfavourable roof proliferation and distribution patterns and unfavourable water transmission properties combine to reduce the efficiency of water use by crops on eroded soils.

In spite of above observations and estimations one should not be oblivious to the complex nature of erosion-productivity relationship. Erosion reduces productivity so slowly that the reduction may not be recognised until land is no longer economically suitable for growing crops. Moreover, improved technology often marks the reduction in productivity (NSESPRPC, 1981). Erosion induced loss in productivity is always non-linear and restorations is very difficult. Till date the research in these areas have not been satisfactory because, first, such experiments are costly and time consuming and secondly, since crop production has been adequate in the past, there has been little incentive for investment in this type of research. A fresh outlook is really needed in this direction.

6.1.2 Other On-Site Impacts of Soil Erosion

In general calculation and impact assessment of soil erosion, only productivity related losses are accounted under on-site category. But there are a number of potential on-farm non-productivity losses due to potential (Crosson, 1985; Pimentel, 1987), which are left unaccounted. They are of course very difficult to calculate and availability of data is scanty. Extent and presence of these types of impacts are dependent on-site specific conditions, *i.e.* soil types, topographic, climatic and cropping pattern. But these impacts must be taken into account. These impacts of excessive soil erosion are:

(a) Lost nutrients, resulting in yield reductions or annual replacement costs.
(b) Lost chemicals, resulting in yield losses due to increased pest pressure.
(c) Emergency rescue or replacement expenses to address post problems on fields where pesticides are washed off or displaced.
(d) The non-uniform distribution of chemicals resulting from soil movements and washing. This may result in yield losses.
(e) Damage to adjacent, incompatible crops due to washed off and deposited chemicals from other fields.
(f) Seed, fertilizer, chemical, machinery and labour costs due to replanting either parts of eroded fields or parts of fields subjected to soil deposition.
(g) Yield reductions due to late replanting on the aforementioned fields.
(h) Costs due to additional machinery wear and tear from ravage in heavier and possibly rock subsoils.
(i) Labour costs to repair machinery used on eroded fields.
(j) Increased energy costs for machinery due to higher drag coefficients or tillage implements on heavier subsoils.
(k) Loss of parts of fields due to gully erosion or severe gully erosion.
(l) Loss of parts of fields due to depositions of eroded materials at the base of the fields.

(m) Increased costs to clean out drainage and irrigation structure obstructed by eroded soils.
(n) Increased irrigation costs due to reduced water-holding capacity of eroded soils.
(o) Increased maintenance costs of existing conservation practices.
(p) Yield losses due to wind erosion damaging young plants.
(q) Wind erosion, which either damages or causes more frequent replacements of parts of farm machinery. It is also responsible for increased maintenance costs of farm buildings and other structures.
(r) Farm expenses associated with cleaning on-farm road ditches and culverts and bridge repair. Replacement or repair of fences covered by deposited soils is also possible.
(s) Reduced property values and possibly lowers land rental rates due to visibly excessive erosion.

This list is neither complete nor a standard one. There could be many other impacts, which directly or indirectly reduce the welfare of people. Experts from different geographical regions can delete or add a few additional impacts depending upon geomorphologic and hydrological factors. Many of the impacts reduce productivity of the crops and should not be counted if productivity loss has already been incorporated under on-site impacts of soil erosion. But many of these impacts are generally not taken into account while doing cost-benefit analysis of conservation programmes for soil erosion (Nowak, 1987, 1988).

6.2 Off-Site (indirect) Impact of Soil Erosion

An indirect effect of soil erosion has commanded a special place in recent years. Soil fertility and local impact of erosion has been taken into account by farmers but the lost soil carried from the farm ends up in local streams, rivers, canals or irrigation and hydroelectric reservoirs and such disturbances cause heavy downstream damages. These harmful impacts include:

6.2.1 Sedimentation

Quantity of runoff and erosion sediments is the outcome of massive economic activity on the uplands, which cause excessive soil loss. Higher sedimentation lead to the growth in siltation causing damage to dams and reservoirs in many ways. This phenomenon of siltation is prevalent all over the world (Table 6.5).

Table 6.5: Siltation Rates in Selected Countries

Country	*Reservoir*	*Annual Siltation Rate (metric tons)*	*Time to Fill with Salt*
Egypt	Aswan High Dam	1,39,00,000	100
Pakistan	Mangla	3,70,000	75
Philippines	Ambuklao	5,800	32
Tanzania	Mathumbulu	19,800	30
Tanzania	Kisongo	3,400	15

Source: El-Swaify and Dangler, 1982.

For Mangla reservoir in Pakistan it was anticipated that the expected life of the reservoir will be around 100 years but the excessive silt has reduced it by around 25 years. Similarly, for Philippines, the anticipated life of Ambuklao dam has come down from 60 years to 32 years due to heavy siltation (USAID, 1979).

For India the phenomenon of siltation due to soil erosion has been even more pronounced (Table 6.6).

Table 6.6: Kilograms of Wheat and Rice Produced per Kilogramme of Nitrogen Supplied as Fertilizers in a Few Reservoirs

Reservoir	*Annual Siltation Rate (in acre-feet)*		*B/A Ratio*
	Assumed A	*Observed B*	
Bhakra (Punjab)	23,000	33,475	1.46
Tungbhadra (Karnataka)	9,796	41,058	4.19
Maithon	684	5,980	8.74
Mayurakshi	538	2,000	3.72
Nizam Sagar	530	8,725	16.46
Ranaganga	1,089	4,366	4.01
Ukai	7,448	21,758	2.92

Source: EL-Swaify and Dangler, 1982.

Raising the concern, B.B. Vohra, then Chairman, National

Committee on Environmental Planning, observed that: "premature siltation of our 50,000 odd ponds and of the 487 reservoirs of our major and medium irrigation and multipurpose projects on which the community has invested over 100 billion rupees during the last three decades is a particularly serious matter". He further states that "... the loss is irreplaceable and will harm the potential of irrigation, electricity and flood control that these storages represent" (Vohra, 1982). At a broader level, Gupta (1980) in his study finds that for 21 major reservoirs in India the annual rate of siltation from a unit was assumed at the beginning of the project. Silts in the reservoirs not only reduces their life but it causes the manufacturing of the irrigation system also. Koshi canal system is an example (Sinha, 1984).

6.2.2 Floods

Siltation of streams and channels reduce the depth and water carrying capacity and increase their width. Satellite images have shown that some of the Himalayan torrents and streams have widened by 100 per cent and rivers by 36 per cent over a period of 7 years (Gupta, 1981). Streams and channels overflow and cause floods bringing havoc on crops, people, livestock and property. This inflicts a very high cost on the local people and the government every year. Between 1953 and 1987 annually, the floods damaged around 7.66 mha areas affecting 31-84 million people. Damages done by floods have been very high for crops (Annual average Rs. 367.79 crores), Houses (Rs. 112.62 crores) and 8 other public utilities (Rs. 287.67 crores) (CSE, 1991).

Power Generation: The impact of soil erosion and siltation has been significant on power generation capability of hydropower projects. Although quantified data for power loss due to siltation is hardly available, logical deduction and experiences of soil scientists eminently suggest that the loss has been significant.

Inland Water Transport: Siltation of streams and rivers reduces their depth and results in restriction or complete loss of capacity to support inland water transport system. According to an estimate, if this erosion continues uninterrupted, the capacity of the Panama Canal to handle shipping will be greatly reduced by the end of this century, forcing many ocean-going freights that have relied in this to shortcut to make the trip via Cape Horn (Woodsworth quoted in Brown & Wolf, 1984).

6.2.3 Inland Fish Production

Siltation of tanks, streams and rivers influences their depth, flow and temperature affecting the breeding and growth of fish adversely. However, no direct evidence of an adverse impact of soil erosion in inland fish and fish-breeding grounds has been reported so far (Khooshoo and Tejwani, 1993).

Off-site impact of soil erosion causes water pollution making it less potable for human population. It also hampers recreation infrastructure etc. However there are some favourable impacts (off-site) of soil erosion, these benign impact of siltation arising out of erosion should not be ignored along with its malign impact. The example of benign impact has been the productivity of soils in low-lying areas of Egypt (Gift of Nile).

7

Economic Estimates of Soil Erosion
Methodology and Empirical Evidences

On-site and off-site effects of soil erosion are distinct and economically significant. All malign (or benign if any) effects need their valuation in concrete economic terms. Valuation of natural resources, like land, water, forest, air and biodiversity has caught attention of a large number of economists. Although, tradition of estimating economic values of natural resources through market and non-market methods is more than fifty years old. Valuation efforts have received an accelerated momentum only recently (Arrow *et al.*, 1993; Hanemann, 1994; Ridker, 1967; V.K. Smith, 1986; 1993, 1996; Bishop, 1982; Eckstein, 1961; Freeman, 1979; Kneese, 1997; Krutilla, 1967; Stocking, 1986; Georgion *et al.,* 1998). Behind these valuations of natural world, a clear-cut mandate from well-established economists belonging to mainstream thinking has been coming (Samuelson, 1983; Leontief, 1936; Knight, 1924; Boulding, 1966, 1969; Dasgupta, 1993; Buchanan, 1987).

Here the purpose is neither to review these works nor to go for debate on valuation issues but to make a moderate attempt to highlight the relevant valuation techniques for impacts of soil erosion. It will be presumed that economic valuation is needed for achieving an efficient decision making process and sustainable economic development. Additionally, the valuation techniques will primarily be addressed to the on-site impacts of soil erosion.

7.1 Valuation Techniques for Soil Erosion

Valuation of soil erosion has been done mainly to facilitate the cost-benefit analysis of soil conservation projects (Bojo, 1992, 1996;

Magrath and Arens, 1989). Following valuation methods are generally used for estimating economic implications of soil erosion.

7.1.1 Defensive Expenditure/Avoidance Cost

The defensive expenditure approach values the damage caused by soil erosion according to the costs, which economic agents, are willing to incur to prevent the damage. Economic agents *i.e.* consumers or producers would like to allocate resources to avoid the damage only if they are convinced that the proposed benefits of avoidance will exceed the costs of bearing it. Here the proposed defensive expenditure can be taken as the willingness to pay for that quality or quantity of soil. Here the cost incurred in preventing the erosion by soil conservation projects, drainage system or irrigation schemes, and similar preventive measures will be taken as the cost of that erosion. This will include both capital and recurrent elements of expenditure. In a study by Kim and Dixon (1986), it was discovered that lowland rice farmers were willing to incur the costs of constructing water diversion channel as a defensive measure against siltation and other productivity damage caused by upland soil erosion in Korea. The defensive expenditure incurred were taken as the minimum estimate of the damage costs inflicted on rice farmers by upland farmers.

The defensive expenditure always gives the lower estimates of the costs borne (James, 1994). In real world undesirable impacts of soil erosion cannot be avoided by averting behaviour in its entirety. It may not capture the true and total decline in the utility of the economic agents. Moreover, averting behaviour can involve costs of time and other resources. A comprehensive analysis of all possible expenditures must precede this approach. In some instances, averting behaviour may yield secondary benefits other than those resulting from damage mitigation. Therefore, the defensive expenditure approach for estimating costs of soil erosion is less readily used.

7.1.2 Lost Production/Productivity Loss

Soil erosion affects the agricultural output. So, crop yields or other output is estimated for the non-degraded and degraded soil and then they are priced. The difference measures the value of lost production. The important issue in productivity loss approach is the requirement of identifying the "with and without" productivity effects. The term "with and without" should not be confused with "before and after". The magnitude of productivity changes is found by comparing the time-stream of output that would occur with and without the soil erosion effects. Here projections or predictions must be made of future levels of output in both situations. Only the productivity changes directly attributable to the erosion are relevant. The two situations, with and without erosion, are assessed by normal methods of farm economics.

Soil erosion in upland areas is predicted to lead to declining crop productivity over a period of time. In the figure, it is shown through PQ time. A favourable change in the farm practices (or conservation efforts), which involves no additional costs, thwarts the decline in crop output, as shown by the line PR. The economic benefits of the improvement programme crop can be measured as the economic value of the savings in output, or the shaded areas in the Fig. 7.1.

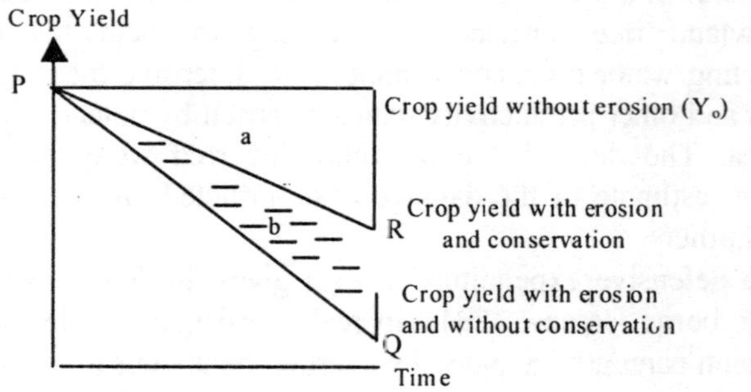

Fig. 7.1: Model of Impacts of Erosion and Conservation on Crop Yield

Here the net benefit increases over time, even though the programme fails to reverse the downward trend in productivity. The total value of the changes in farming practices can be measured as the present value of the change in net benefits up to a prescribed planning horizon such as H.

This method is widely used for estimating damage due to erosion, but erosion-productivity relationship is quite complex. A dose-response kind of estimate is merely an approximation and not a precise estimate.

7.1.3 The Replacement/Repair Cost Approach

This approach relies upon the incremental expenditure incurred on the replacement, maintenance or repair of soil health. The erosion of soil should exclusively be attributed to man-made factors and excludes natural erosion. Generally, farmers put many efforts to avoid losses in production and they do it through higher use of fertilizer inputs. For the same yield level, the additional fertilizer needed is a measure of the cost of degradation. This can be treated by estimating the quantity of nitrogen, phosphorus and potassium removed by eroded soil and then the cost of degradation is valued at the cost of replacing these nutrients by fertilizer. Michael Stocking (1986) has used this approach for estimating the cost of soil erosion for Zimbabwe. Another example of this approach is a study conducted by Kim and Dixon (1986) for examining the cost of soil lost in upland Korea.

This approach is based on the assumption that the replacement/ repair cost is at least as high as the value of the asset that is damaged; otherwise it would clearly not be economic to replace the asset (soil). This approach is applicable only where the magnitude of the damage can be measured and where the replacement cost can be estimated. But one should not be oblivious to the fact that topsoils are complex ecosystem exhibiting all the characteristics of uncertainty and vagueness and the cost of replacing fertilizers cannot capture the value lost organic components and useful soil biodiversity. This method, at the most can estimate the conservative side of the loss.

7.1.4 The Property Value Approach

This approach attempts to measure environmental damage costs or environmental improvement benefits by tracing the effect of environmental quality on property prices. It is further based on the economic concept that the value of a property is directly related to the present value of the expected stream of benefits to be derived from that property. If there are two pieces of agricultural land having similar attributes except the degree of erosion, then the difference in prices can be attributed to the erosion factor.

$$P = f(\alpha_1, \alpha_2, \alpha_3, ... \alpha_n, E)$$

If P be the price of the land, α_1, ... α_n the attributes of the land like location, natural features etc. and E is the level of erosion. Cross-section regression analysis can yield the price of the cropland and cost of the erosion. In spite of being a technically robust method this approach has little relevance for developing countries where efficient markets for agriculture land hardly exist.

7.1.5 Restoration or Reclamation Cost

It is the cost of restoring the soil to its former productive state. All previous approaches assess the cost of soil loss incompletely. For any level of soil loss, it is not adequate to value the cost of installing soil conservation measures, followed by improved land management, for one is still working with the deputed, shallowover soil. For the true restoration of soil health, besides replacing the nutrient, organic matter should also be restored which may be possible by allotting part of the land to a green manure crop, foregoing production and finally replacing the soil.

For soil replacement, one can buy the soil from a horticulture supplier and put a market value on it but it will create other environmental concerns. Alternatively, one can take the land out of the production until the natural process of weathering restores it. The rate of weathering will depend upon local gemorphological condition but it has been estimated that a maximum of 500 mm thick soil layer can be restored in 1000 years (Saunders and Young, 1983). Soil scientists put the rate of soil formation at one inch in 30-300 years or about 0.1-10 mm per year (Schertz, 1983). Presuming natural erosion to be very slow, even the fastest of the rates of soil formation would mean putting the land fallow for 50 years to restore the lost 5 cm and hence forfeiting the future income stream of lost production. This type of cost estimation may be very high or sometimes unacceptable but it gives a measure of the true resource loss.

Valuation methodologies for soil erosion vary in their application and acceptability depending upon the data available and the purpose of the researcher. If all the methods can be applied, then in a competitive setup the minimum among those should be acceptable (Parikh and Parikh, 1997). Parikh *et al.* recommend that:

Cost of land degradation = Min [(Avoidance cost), (Restoration cost), (Loss in Productive Asset Value)]

7.2 Empirical Evidences

The Food and Agriculture Organisation (FAO) has been involved and been recommending the economic valuation (cost) of land degradation under Soil Conservation Programme of FAO. A case study of Zimbabwe was undertaken for cost estimation of soil erosion in 1986. The principal investigator of the project, Michael Stocking adopted the replacement of nutrient lost due to erosion. He found that on an average 1.6 m tons of nitrogen, 15.6 m tons of organic matter and 0.24 m tons of phosphorous are lost annually by Zimbabwe. Monetary values of this physical loss were calculated by the prices of equivalent fertilizers. It came out as US$ 1.5 billion per annum [at 1985 market fertilizer prices and rates of exchange (Stocking, 1986) for the arable land]. The financial cost was estimated to be around US$ 150 m.

The Winand Staring Centre of the Netherlands undertook another comprehensive study, supported by FAO. Their analysis measured the net removal of the macronutrient nitrogen, phosphorous and potassium from the rootable soil layer using production figures for 1983 and projected figures for 2000 provided by FAO by crop and country. This study group calculated the cost of erosion in arable area to around US$ 164 million.

Both these studies ignored secondary and other off-site costs due to the lack of data in extremely heterogeneous situations.

For developing countries, the experimental data showing productivity—erosion relationship is virtually absent. Experiments and subsequent result of temperate regions are no longer applicable for tropical countries. However, several studies of third world show a very high cost of soil erosion. A study conducted by Magrath and Arens (1989) in Java showed the on-site and off-site costs of declining soil productivity at US$ 340 million to US$ 405 million or 0.5 per cent of total GDP. Bishop (1990) in a study for the World Bank estimated the costs of soil erosion and found a link between soil loss and crop yield. He calculated the economic losses to be about 4 per cent to 26 per cent of net farm income for Malwai.

In another study for the World Bank, Bishop and Allen (1984) estimated the on-site costs of erosion in Mali. Their study estimates a mean current rate of soil erosion of 20t/ha/y on gross arable land for entire Mali. The on-site costs of soil erosion are expressed in terms of reduced crop yields, resulting in mean annual yield losses between

4 per cent and 25 per cent. They estimate the gross annual losses between 0.5 per cent and 3.1 per cent of 1988 GDP for Mali. For Zimbabwe again, Groh (1993) estimates the economic costs of soil erosion to small-holder farmers through the productivity loss approach. The methodology uses the Soil Loss Estimation Method for Southern Africa (SLEMSA) for field level estimation, and a GIS aggregation for aggregate costs. Off-site costs in small dams have also been estimated. Grohs has found that a high 3 per cent yield decline per 1 centimetre of soil lost; small holder form income losses for 1989 or $4.4 million and $1.7 million losses for the large-scale sector. These estimates are lower than those given by Stocking (1986).

For India, Parikh and Ghosh (1991) have done valuation of soil loss from resource accounting point of view. Their valuation is primarily based on loss of productivity and does not exclusively attribute to man-made erosion.

In another pathbreaking study, two World Bank staffers, Brandon and Homman (1995) have estimated the economy wise cost of soil erosion. They also depend upon FAO studies of 1994 (cited above) in absence of data but they go much beyond FAO's studies. They have estimated the productivity losses due to land degradations between US$1.5 billion to US$ 2.4 billion. The estimated loss is roughly evenly divided between dry and humid zones. Restoration costs of eroded soil have been estimated in the range of US$ 30-50 billion by taking 5 per cent rate of discount for a long period for which production will be lost. Thus we see that the cost of soil erosion is very high but may help in rationalizing the cost of conservation programmes.

8

Case Study of Doon Valley

8.1 Study Area: Location and Geophysical Environment

Doon valley is situated in Uttaranchal state of India and southwest of Himalaya-Siwalik region and lies between 30°-30°28′ N latitudes and 77°32′-78°20′ E longitudes. The valley is bounded in the northeast by Lesser Himalayan belt and by Siwalik ranges on the south-west. Two important rivers of north India Yamuna and Ganga demarcate the south-eastern and north-western boundaries of Doon valley. Maximum east-west length of the valley is approximately 70 km and the width varies from about 2.5 km in the extreme west, 25 km in the central part to nearly 45 km in the east. The terrain is undulating and the altitude range from 315 m to 2,500 m. Doon valley referred to in the study coincides with the boundaries of Dehradun tehsil.

Doon valley described here conforms to the Dehradun tehsil of Dehradun district having a geographical area of 2,090 sq. km. Dehradun tehsil has been subdivided in four administrative blocks, *viz.* Vikaspur, Sahaspur, Rampur and Doiwala (Fig. 8.1).

Dehradun, the largest centre among all the urban settlements of hill districts of Uttar Pradesh and Mussoorie, also known as Queen of Hills, have centres of all developmental activities in the region. Availability of infrastructure in these two urban settlements has given impetus to the development of regional linkages and played a vital role in the development of the region. The natural beauty and climate of Doon valley have also provided the potential for a prosperous tourist industry. In addition, valley's location as the gateway to the famous pilgrimage centres of Kedarnath, Badrinath, Gangotri, Yamunotri, etc. The combination of health, pilgrimage and aesthetics based tourism has made Doon valley a popular tourist resort

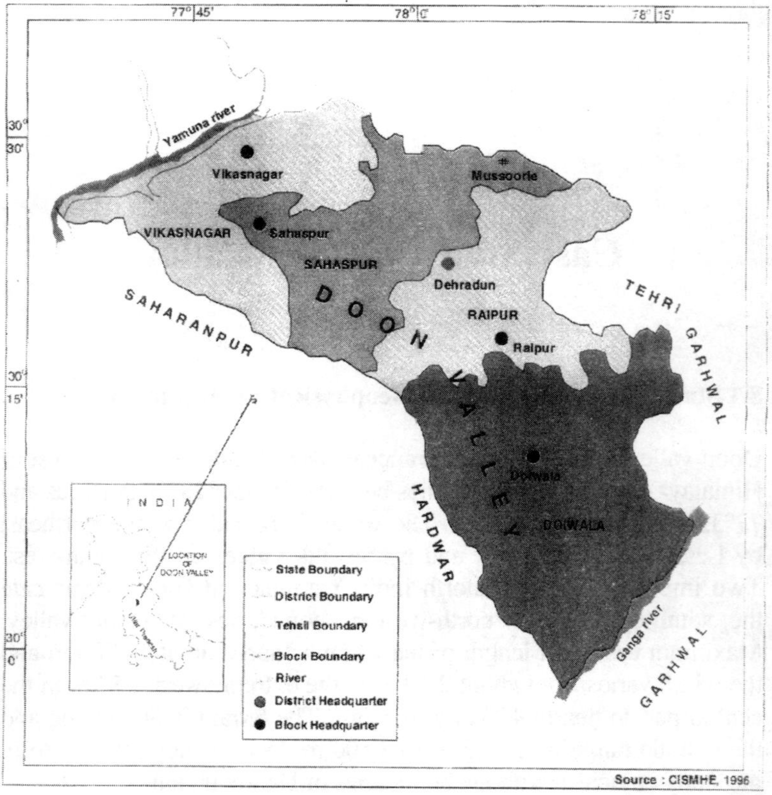

Fig. 8.1: Study Region

contributing substantially to the economy of the valley. The well-known towns of the valley are Dehradun, Rishikesh, Raiwala, Doiwala, Dakpathar and Sahaspur along with Mussoorie and Dhanaulti hill resorts.

Doon valley ecosystem is comprised of two distinct sub-catchments, one formed by the drainage basin discharging into the river Ganga, south of Rishikesh and the other by the drainage basin discharging into the river Yamuna near Paonta Sahib. The areas draining into Yamuna and Ganga rivers cover nearly 37.83 per cent and 62.17 per cent, respectively. Doon valley thus forms a sub-catchment for the Ganga-Yamuna river system, which is lifeline to the northern part of the Indian subcontinent.

8.1.1 Physiography

The Lesser Himalayan range, which forms the northern boundary of the valley, is part of the Great Himalayan ranges. The Siwalik range, forming the southern boundary of the valley are an alluvial formation younger than the Himalaya since they are formed by the debris swept from that mountain range. The Siwaliks exhibit a stiff face towards the plains while a long and gentle slope meets the foot of the Himalayan ranges forming a longitudinal shallow valley. These valleys or longitudinal depressions formed between the Siwaliks and the Himalayan are generally called 'Duns'. They are not continuous but are cut through by the passage of streams that drain the interior mountains. In some places the Duns disappear with the merging of the Siwaliks and the Lesser Himalaya. The lower part of those 'Duns' are generally covered with a deposit of boulders so that the floor of the valley is considerably higher than the level of the plains beyond the Siwaliks.

The north-eastern slopes of the Siwaliks are very steep and rugged in the upper portions, but lower down they have gentle gradient. These are dissected by a large number of '*raus*' (torrents) draining into Asan, Suswa and Song rivers. Drainage on Siwalik slope is braided. The valley proper, on both sides of Asan River and Rispana, Song and Jakhan rivers is more or less flat or has gentle slopes. The forests on both banks of the Asan and Song rivers are flat or gently sloping areas often cut-up by deep nalas.

The slopes of Outer Himalayan ranges are characterised by gentle to moderately flat tops with steep slopes and deep valleys.

Doon valley, although apparently a single valley, actually comprises two shallow valleys, the Western Doon and Eastern Doon. A low watershed of about 610 m separates them to 762 m altitude running from Mohand Pass in the southwest through Majra, Garhi, and Rajpur to Landour in Mussoorie. The Western Doon forms the portion of drainage basin of Yamuna whereas the Eastern Doon forms the drainage basin of the river Ganga.

8.1.2 Geology

Geologically, Doon valley is a synclinal trough within the Siwalik system. The limbs of the syncline consist of the middle and upper Siwalik rocks followed by the northerly dipping pre-tertiary formations of Lesser Himalaya in the north. On basement of Siwalik rocks, Dun gravels are deposited. Stratigraphic succession of Doon valley is given in Table 8.1.

Table 8.1: Stratigraphic Succession of Doon Valley

Geological Unit	*Lithology*	*Age*
River Terraces Flood Plain	Clay, silt, sand, gravel, pebble, etc.	Recent
Sub-recent Fan Terraces	Silt, clay, sand and gravel etc.	Sub-recent
Doon Fan Gravel	Loose stratified beds pebble, clay and sand	Holocene
Old Doon Gravel	Massive pebble, boulders, conglomerate with pockets of red clay	Late Pleistocene
Upper Siwalik	Pebble bed, conglomerate and clay bed.	Pliocene to Pleistocene
Middle Siwalik	Sandstone and clay bed	Miocene to Pleistocene
Main Boundary Thrust		
Tal	Sandstone shales	Lower Palaeozoic
Krol	Limestone	Lower Palaeozoic
Infra-Krol	Shales, slates and quartzite	Lower Palaeozoic
Blaini	Boulder beds	Pre-cambrian
Nagthat	Quartzite	Pre-cambrian
Chandpur	Phyllite and slate	Pre-cambrian
Damta	Quartzite, sandstone and Greywacke	Pre-cambrian

Source: CISMHE, 1994.

Doon valley is classified into the following geomorphic regions:

(i) Siwalik structural denudation hills,
(ii) Dun gravel,
(iii) Piedmont colluvial tract, and
(iv) Lesser Himalaya structural hills.

(i) The Siwalik System

In the Siwalik hills, three lithological divisions can be easily identified into lower, middle and upper. The composition of the Siwalik deposits shows that these are alluvial detritus derived from the sub-aerial waste of the mountains, swept down by numerous rivers and streams and deposited in the long narrow depression at their foot. The Siwalik sediments have been folded and elevated during the latest Himalayan upheaval.

The southern limit of Doon valley is defined by middle and upper Siwaliks. The middle Siwalik is a group of coarse incaceous sandstone and subordinate shales. The upper Siwalik consists of coarse conglomerates and massive beds of sandstone. These are richly fossiliferous.

The weathering of the Siwalik rocks has been proceeding at an extraordinarily rapid rate since their deposition. This has given rise to abrupt forms of topography. Broad longitudinal strike valleys known as 'Duns' separate gigantic escarpments and dip slopes. The strike is remarkably constant in a NW-SE direction, with only brief local swerves. It is in conformity with the axes of ridges.

(ii) Doon Gravel

The asymmetrical synclical valley is filled with the sediments (Dun gravel) derived from pre-tertiary and Siwalik rocks. An isolated hill named Nagsidh is present in the eastern part of valley. It is composed of big broken boulders and pebbles with pockets of clay. These are known as old Doon gravel. Loose stratified beds of pebbles, sand, clay and gravel forming sub-recent fans and terraces occupy rest of the area of Doon valley. The Dun gravels are of late Pleistocene/ Holocene age.

The drainage pattern in the valley where Dun gravel is deposited is quite significant. There is a main water divide trending more or less N-S and passing through the city of Dehradun. It separates Ganga and Yamuna river system. The eastern part of the valley is drained by Suswa, Song, Jakhan and Kans *rau* in Ganga whereas Asan, Sudhonwala *rau*, Chorkhala *rau* and Sitla *rau* traverse through western part of the valley before joining Yamuna.

(iii) Piedmont Colluvial Tract

The hill slopes bordering Himalayan side are covered by piedmont colluvial tract. This material has been derived from Lesser Himalayan rocks. Hydrologically, piedmont Dun gravels are considered ideally disposed to yield some good quantities of water supplies.

(iv) Lesser Himalayan Structural Hills

Towards north, the Siwaliks directly come in contact with the rocks of Krol formation. The oldest litho-unit of the Lesser Himalayan succession in the present area is the Chandpur formation overlain by the Pre-Cambrian to Lower Palaeozoic Nagthat, Blaini, Infra-Krol, Krol and Tal formations successively, which form the northerly dipping southern limb of Mussoorie syncline. The composition of the main geological formations is given below:

(a) *Chandpur Formation*: It consists of phyllites and quartzite and forms subdued hills.
(b) *Nagthat Formation*: It is mainly quartzitic sandstone with some pebble beds. The quartzitic hard sandstone often form cliffs.
(c) *Blaini Formation*: It is composed of boulders with slates. The pebbles are of pink quartzite, slate and limestone.
(d) *Infra-Krol Formation*: It is made up of highly incompetent rocks, chiefly shales and slates with quartzite bands.
(e) *Krol Formation*: It consists of limestone and shales at the base, passing gradually to dolomite towards top with inter-bedded shales.

Limestone is widely quarried for the requirements of chemical industry and the preparation of lime. The town of Mussoorie is situated on this formation. To the East of Mussoorie, Krol formations are overlain by Tal. Tal formations consist of dark sandstones and shales. Limestone is generally a poor supporter of vegetation and this area is relatively devoid of vegetation. Loose and incompetent rocks such as Infra-krols and Chandpur are infested by landslides and are prone to erosion.

8.1.2.1 The Main Boundary Fault

The Main Boundary Fault is a noteworthy tectonic feature in the geology of the Siwalik foothill zone all along its length from Indus to the Brahmaputra. In Doon area it is superimposed by the Krol Thrust separating Siwaliks and rocks of Krol nappe. The Main Boundary Fault appears to be neotectonically active as revealed by changes in the river courses and geodetic observations. There are two prominent landslides near Nalota nala and Kalagarh in vicinity of Main Boundary Fault.

8.1.2.2 The Yamuna Linear

The Yamuna linear, which displaces the Siwalik range around Paonta Sahib and shows up through the alluvium in the foothill zone is observed to extend further deep into the pre-tertiary sediments roughly corresponding to the alignment of the Tons valley. This has also been described as Yamuna Fault or Paonta Fault. This feature has also influenced drainage and flow direction of river Yamuna and its tributaries.

8.1.2.3 Drainage

Asan river mainly drains the forested catchment of western Dun while in Eastern Dun Song and Suswa rivers drain these. Major tributaries of San rise in the Mussoorie ridge of Himalayan range. These are Tons, Nun, Nim, Gulati, Suarana, Chorkhala and Sitla *raus*. A large number of *raus* rising in the northern slopes of Siwaliks discharge in the Asan river from the south.

Suswa rises very nearly opposite Asan river to the left of Saharanpur-Mussoorie highway and flows in a southeasterly direction to discharge. Song rises in the Saklana area of Tehri district near Surkanda peak on Mussoorie-Chamba ridge. It flows westwards along the northern face of Ladwakot-Paled-Maidan ridge and forms the boundary between Tehri and Dehradun Forest Divisions. It then enters Doon valley at Maldeota wherefrom it flows almost to the south up to its confluence with the Suswa in Banbaha block. From there it flows in a south-eastern direction till it drains into Ganga near Satyanarain. The important tributaries of Suswa are Bindal *nala* and Rispana *rau*, whereas Bandal, Reh and Jakhan are the important tributaries of Song. Song river drains Raipur, Ladpur and Dwara blocks while Mohbewala, Asarori, Mohammadpur forest blocks form the drainage basin of Suswa river.

The *raus* carry immense volume of water during monsoon, which continuously undermines the containing banks, and as a result they have very wide and shallow beds. Most of the *raus* are dry during the greater part of the year, but are torrential during the monsoon carrying large volume of boulders and loose shingles down their beds.

The valley also has several natural springs at different altitudes. The well-known spring, Sahasradhara, is a popular tourist bathing spot due to the present of sulphur in its water.

8.1.3 Soils

Due to wide variation in topography, intensity of erosion, parent material and other factors, the soils show wide variation in many characteristics specially texture, depth, stoniness, colour, drainage, moisture status, organic matter content and cation exchange capacity. In fact, soils are the products of original geological formations hence three geological zones exhibit difference in their soil types. The detailed information on soil types has been given in (Fig. 8.2). However, broad categories of soils in different geological belts of Doon valley are described below.

8.1.3.1 Lesser Himalayan Belt

The Himalayan belt contains only a small bit of the tract especially in Nahan formation of the sub-Himalayan series. It consists principally of loose sandstone and hence soil is arenaceous in nature. It is well suited to forest growth by providing a good foothold to trees even where there is no soil.

8.1.3.2 Siwalik Belt

The Siwalik belt consists of sandstone and conglomerates, which are often soft and friable. Bands of clay are interceded with these sandstones and conglomerates, giving cohesion to the soil and thereby improving its physical qualities. The soil resulting from the Siwalik sandstone is a sandy-loam, with sometimes a large proportion of clay. In higher regions with steep slopes the soil is very shallow and dry. Consequently the vegetation as a whole is more xerophytic. It is, however, deep on northerly aspects and in lower slopes. The best soils (which are also most eminently suited for *Sal*) occur on the lowest slopes where the proportion of clay is large and the drainage is adequate. But such conditions do not prevail over large areas.

8.1.3.3 The Boulder Belt of the Valley

The boulder belt is characterised by underlying beds of boulders and shingles of great thickness. Thin bands of clay are also intruded in these deposits causing swamps where they crop out. The soil, overlying the boulder mass, is very variable both in depth and richness and is altogether absent in *rau* beds. It is generally, sandy and poor with varying proportions of clay locally. Because of great thickness of underlying boulders, the water table is at a great depth.

Soils of Lachhiwala, in the south of the valley in Eastern Doon are generally silty loam to silty clay in texture and acidic to near neutral in reaction containing a high amount of organic carbon. The higher percentage of base saturation indicates adequate level of available nutrients in them. Non-calcareous soils are characteristic of Thano block where too the soils are acidic to neutral and the texture varies from loam to loamy sand. These soils support *Sal* as the main vegetation.

The soils of Jhajhra in Western Doon differ considerably due to physiography and relief. The soils are highly eroded. Soil cover is thin; texture is very coarse and nutrient status very poor. The soils, however, at the lower portion of gentle slopes are fine loamy to coarse loamy in texture and acidic in nature. The soils of Barkot range in the Golatappar area in Eastern Doon do not exhibit abrupt change in texture. These are acidic to near neutral in nature and fine loamy in texture. These support *sal* as well as miscellaneous vegetation.

8.1.4 Climate

The climate of Dehradun is more temperate and humid than that of the adjoining districts, the maximum and minimum day and night temperatures being 3 to 6°C lower throughout the year. Only the months of May and June are hot though they are seldom oppressive.

8.1.4.1 Temperature

The temperature as recorded in the Forest Research Institute Observatory during the last decade varied from 3°C in December to about 38.2° in May. The highest temperature recorded at the Survey of India Observatory at Dehradun since 1868 was on June 7, 1923 when it rose to 43.65°C undershade. The lowest temperature was on January 16, 1946 and February 28, 1956 when it fell to –1.1°C. Other exceptional frost year was in 1930 when the temperature fell to 0°C on December 30.

8.1.4.2 Frost

Frost is quite common on winter nights. Severe frost occurs in January and February and cause damage to young *sal* regeneration. There are many places especially in the low grasslands and within two miles on each side of the Suswa and Song rivers, where the night temperature during October to April are several degrees lower than those at Dehradun. In these places frost occurs as early as the first week of November.

8.1.4.3 Winds

The prevailing winds come from the west and are generally of moderate velocity, except when severe storms occur during the hot weather, in April and May. In the evening there is always a light breeze coming down the hills generally from the north. The hot '*loo*' rarely manages to cross the Siwalik ridge. Nightly dews prevail uninterrupted throughout the year except for April, May and part of June.

8.1.4.4 Rainfall

The monsoon generally breaks out by the end of June and the rains continue until about the middle of September, the heaviest falls normally occurring in July and August. Occasional showers may continue in September and October after which, there is usually little rain until about the end of December. Some rain usually falls in January and February after which there is very little rain except for occasional thunderstorms until the break of the monsoon. Few months, however, pass without any rain at all; however, bad drought years are rare in Doon valley.

The average annual rainfall in the valley varies considerably. It varies between 2,200 m and 2,500 m in Mussoorie in the Outer Himalayan range, between 2,200 and 2,400 mm in Dehradun proper and Rajpur receiving the maximum rain in the valley varying between 2,800 and 3,100 mm. The wettest month in Dehradun was August, in the year 1885 when it received 1637.8 mm of rainfall in the month. The wettest year was, however, 1884 when annual rainfall of 3,118.9 mm was recorded in the valley.

8.1.5 Land-use/Land cover

The forests are predominant land-use in the valley covering approximately 56 per cent of the valley (based upon satellite data of January, 1994 FCC) (Fig. 8.3). Cultivation is the second major land-use in the valley (27 per cent). About 19 per cent of this cultivation is practised on hill slopes and rest being done on gentle slopes or in plains. Sal forests are predominant forest type in the valley, Siwalik belt, valley proper and residual hills. The southern slopes of Lesser Himalayan region as covered with pine forests at lower altitudes and banj-oak forests at higher altitudes, found mixed at places with rhododendron, fir and pine. The gravelly alluvium of streams and rivers is covered with Khair-sissoo forests. The degraded forests cover as much as 29 per cent area of the valley.

The distinctive topographic, climatic and edaphic factors of Doon valley, characterised by their vicinity to tropics in the south and Himalayan highlands in the north, has bestowed the region with rich biological diversity.

8.1.6 Mineral Resources

Limestone, gypsum, phosphoric ore and marble are amongst important economic minerals besides various types of building stones. The bedded limestone of the Upper Krol Formation is of cement grade. Till recently this limestone was mined for manufacture of lime chemicals and cement. These mines were closed some year's back as unplanned exploitation had adversely influenced ecosystem of the region.

There are a few workable pockets of gypsum within the Krols near Sahasradhara and Chiphaldi. The Krol-Tal contact in the Mussoorie syncline is marked by the presence of 3 to 3.5 m thick phosphoric belt which has been found to be a most suitable natural fertilizer for acidic soils. The company Pyrite Phosphote Chemicals Ltd. (PPCL) is mining these deposits at Maldeota and Durmala. In addition there are some minor occurrences of minerals like barite, galena, marble and lignite. However, economic importance of these minerals is yet to be ascertained.

8.1.7 Limestone Quarrying and Doon Valley

The limestone deposits of Doon valley lie in a fragile and ecologically sensitive region. The limestone deposit has an important role in the stabilization of land and water resources in view of the steep slope of the hills in which the deposits are located, the intrinsic geological instability in the deposit belt, since it lies in an active tectonic zone, the overlap of the deposit with catchments of important water resources, the high rainfall of the region which can accentuate the instability of the hill slopes and the aquifer role of the limestone deposit. Limestone quarrying has been undertaken in this fragile area without any serious considerations of these ecological functions of the belt, which are of vital economic value. Serious ecological impact of land and water instability has been generated in the region as a consequence. The destruction of land and water resources has in turn damaged the prospects of economic activity like agriculture, animal husbandry, tourism and knowledge industry.

Limestone deposits in Doon valley have been the centre of controversy over resource utilization at both the local as well as national levels for more than a decade. The beginning of the extraction of Dehradun limestone was for modest purposes like the preparation of lime for construction and whitewash. It later found use in the North Indian sugar and textiles industries. Finally, due to its chemical purity (over 99 per cent) it is being used in the steel and chemical industries in far away places like Renukut and Jamshedpur in Bihar. The resistance to the extraction of limestone from this vulnerable ecosystem first came from local villagers whose agro-pastoral economy was threatened by limestone quarrying. It later came from the State and Union Governments due to the severe ecological impact of the extraction of a mineral that also plays a vital role in the ecological stability of the region.

8.2 Causes of Soil Erosion in the Doon Valley: Socio-Economic Dimension

Scenic beauty and moderate temperate climate has been attracting the people from the plain region of UP, Delhi and Punjab since the Raj

Period. This attraction could become the genesis of unsustainable economic activities causing the erosion of very high order. Following are the reasons for the soil loss in the Doon valley.

8.2.1 Population

In last 100 years, the population of the region has gone up significantly, only 47,757 of population in the whole Dehradun became 1,16,945 in 1872 and reached to 1,78,195 in 1901 (Gazetteer of Dehradun 1995). Since 1901 the population has increased continuously and has crossed 10 lakhs in 1991 (Census 1991).

During 1901-1991 the population of India has increased by 254.99 per cent, for UP the growth rate has been 186.07 per cent and Dehradun it has registered around 478.0 per cent for the same period. As is clear from the table during 1901-1941 the population for Dehradun has not grown very high but after 1951, for every decade the growth has been more than 30 per cent except 1951-1961. These figures are not only for the valley but also for Dehradun district, *i.e.* for its 6 blocks. The blocks of Doon valley, *i.e.* Vikasnagar, Sahaspur, Raipur and Doiwala (Doon valley) experienced the same pattern of population explosion. Contribution of immigration has been quite significant. This explosion has accelerated the demand for biomass, which led to the pressure on agriculture, encroachment into the forest area leading to the land clearing process and agriculture along the slopes. Greater demand for residential complex, road transport and other infrastructural facilities caused damage to the soil. Excessive growth of population in the valley dramatically changed the land-use pattern also.

8.2.2 Changing Land use Pattern

In Doon valley, as the population has exploded, land under agriculture has gone under a change. Forest has been cleared and more and more land has been brought under agricultural use. Official sources of data on land-use pattern especially for forest area is far from perfect.

The methodology of data collection is highly unscientific and unreliable. At tehsil level, data keeper is Land Record Officer who reports to the district statistical officer about the land-use pattern. Information on forest area, barren land, fallow land, permanent pastures, gross sown area emanates from agricultural survey, which is

Table 8.2: Population of India, UP and Dehradun Since 1901

Year	*Population in millions*		
	India	*UP*	*Dehradun*
1	2	3	4
1901	238.40	48.628	0.177 —
1911	252.10 (5.74)	48.155 (-0.97)	0.204 (15.3)
1921	251.30 (-0.30)	46.672 (-3.08)	0.211 (3.6)
1931	279.00 (11.00)	49.780 (6.66)	0.229 (8.5)
1941	274.00 (14.22)	56.535 (13.57)	0.265 (15.6)
1951	361.10 (13.31)	63.220 (11.82)	0.361 (36.1)
1961	439.20 (21.64)	73.755 (16.66)	0.429 (18.6)
1971	548.20 (24.80)	88.341 (19.78)	0.577 (34.6)
1981	685.20 (25.00)	110.862 (25.49)	0761 (31.9)
1991	846.30 (23.51)	139.112 (25.48)	1.025 (34.7)

Source: Census of India, Registrar General and Census Commissioner for India, 1991. Figures in parentheses show the percentage change over the previous decade.

conducted after every five years. This becomes the base information and every year the village Patwari furnishes the information on area under different crops. Area coming under commercial use in the urban suburbs is collected from the local municipal corporation. Against the reported area, adjustment is made and figures for different land-use is arrived. Following Table 8.3 gives the detailed information on land utilisation for Dehradun since 1974-75.

From Table 8.3, out of the reported area forest area has increased from 1,70,712 ha in 1974-75 to 2,23,000 ha in 1991-92, but the trend is not clear. Even this growth in forest area does not comply with the reality and expert's opinion. Classification and measurement of forest area is so conventional and rudimentary that any analysis on the basis of this kind of data will lead to grossly erroneous conclusion. Sudden rise in forest area in year 1977-78 and 1991-92 is difficult to justify.

Under permanent pastures the lower one in 1981-92 and 1982-93 then rising up to 56 ha in 1984-85 is difficult to explain except a statistical anomaly. Under gross sown area, it is the sum total of net sown area and area sown more than once. It is showing increasing trend but inconsistently.

From this table, one cannot rely upon the picture for deforestation in the valley. Forestry data needs expert hand to collect. In the table, total forest is increasing without showing any affronted area.

Table 8.3: Land Utilisation for Dehradun (ha)

Year	*Reported Area*	*Permanent*		*Gross Sown Area*
		Forest	*Pastures*	
1974-75	265797	170712	32	81578
1975-76	266338	171015	34	83385
1976-77	267119	170751	36	84514
1977-78	318328	222456	13	86088
1978-78	316903	219664	15	89198
1979-80	317925	220996	10	80612
1980-81	322785	226086	14	88541
1981-82	312151	215514	4	85245
1982-83	320189	222391	5	87940
1983-84	319247	220056	15	87664
1984-85	315459	220019	56	85493
1985-86	314665	219168	52	87450
1986-87	319347	219176	52	98534
1987-88	315023	219176	54	87561
1988-89	314571	219213	47	84617
1989-90	314899	219811	27	89997
1990-91	315497	219901	99	85560
1991-92	315382	223000	81	91594
1992-93	315672	N.A.	81	86488
1993-94	315032	N.A.	81	87706
1994-95	315751	N.A.	81	87987

Source: *Sankhiyaki Patrika*, Distt. Dehradun, Economic and Statistical Division State Planning Unit, UP, (All figures in hectares).

Water-induced soil erosion is a direct function of the forest cover; therefore, deforested area of Doon valley is needed with utmost sophistication and precision. For this purpose, I have relied upon the satellite data furnished by Centre for Interdisciplinary Studies on Mountain and Hill Environment (CISMHE), University of Delhi. As will be shown latter, data on land-use pattern for Doon Valley generated from different imagery shows that deforestation has been taking place in the valley.

8.2.3 Expansion of Agricultural Activities in the Fragile Sloppy Land

In Dehradun valley, a large number of settlers from other parts of the country (mainly public sector employees and government bureaucrats) have arrived there in the last three decades and they have settled there in the city which has encroached the agricultural areas in the valley. This process has compelled local people to look for agricultural land in the valley. Thus mountain scrubs area has been cultivated. Agricultural land has been developed where the land slope exceeds even 50 per cent slope mark.

Table 8.4: Area Under Hill Cultivation in the Valley

Year	*Area ('000 ha)*	*Per cent Increase (Over Previous Observation)*
1970	5.68	—
1972	6.56	12.31
1974	7.40	12.98
1976	8.10	9.38
1978	9.80	21.00
1980	10.59	7.98
1982	10.82	2.19
1984	11.25	3.98
1986	12.13	7.84
1988	12.80	5.53
1990	12.83	0.18
1992	13.00	1.40
1994	13.00	0.00

Source: CISMHE, 1994 & IIRS, 1980, 1989.

Cultivation on fragile mountain land by marginalised farmers, which causes massive soil erosion especially in underdeveloped tropical countries, is well documented in the literature (Eckholm, 1976; Southgate (1990); Blackie & Brookfield, 1985). In Doon valley also severe erosion has occurred due to cultivation of hill area with much steeper slope and very nascent soil type. Table 8.4 gives the growth of cultivated area in the hills in the valley over last 25 years.

These figures are based on imagery interpretation and have been compiled from CISHME's report (for year 1972, 1974, 1978, 1980, 1982, 1993 and 1994) and rest has been derived from Indian Institute of Remote Sensing's presentation reports. During 1970-74, the growth rate in the cultivation area on the hills has been more than 12 per cent. This growing encroachment is fragile and sloppy land for unscientific farming which size is generally less than even 1 ha has contributed the alarming rate of soil erosion in the valley.

8.2.4 Mining Operations

In Doon valley mining and quarrying activities for limestones and phosphate has been quite prevalent. This process breaks the soil and makes it more prone to the erosion during the monsoon. This process gets further accelerated if post quarrying restoration phase is not properly done. Unscientific mining and non-adoption of proper rehabilitation measures in most of the abandoned mine sites in the valley have created the problems of landslide, denudation, shifting mass of land and rubble, pollution of soil, blockades of roads and reduction in water holding capacity of streams. But the single most important effect of unscientific mining activities is the soil degradation due to mining activities in the Doon valley, which has reduced food production by 28 per cent, water resources by 50 per cent and livestock production by 35 per cent. This is typical the case with the quarrying activities in the valley. The data related to mining area in the valley is not in published form, so a visit to Tehsildar (mining) in Dehradun was made in order to collect this information. Following is the area under mining in the valley.

Over the period of 1965-82, the area under mining has registered a growth rate of 101.25 per cent. Some of these areas fall under Mussoorie District and are inseparable from the area of Doon valley. In 1982, under the court order, mining has been banned in this area but besides some mines of UP State Mineral Board, illegal activities are still going on. Of course, statistics do not show any growth in the mining area of the valley after 1982.

Table 8.5: Mining in Doon Valley

Year	*Area (in acres)*	*Percent Increase (over previous observation)*
1965	1370.19	—
1970	1501.10	8.72
1071	1536.69	2.37
1972	1596.69	3.90
1973	1630.86	2.14
1974	1709.53	4.82
1975	1918.53	12.22
1976	2325.70	21.22
1977	2325.70	0.00
1978	2368.70	1.84
1979	2535.73	7.05
1980	2620.50	3.34
1981	2620.50	0.00
1982	2575.57	5.22

Source: DM, Dehradun, 1998.

8.2.5 Livestock

Livestock has registered a consistent growth over time in Doon valley. Growth in population especially that of rural folks have propelled them to depend upon the livestock. In absence of proper skill and education, rural population by and large has not been able to engage itself in the modern industries. Now growth in the number of livestock directly puts pressure on grassland, which leads to overgrazing, and hence soil erosion. A heavy dependence on livestock be it goats, sheep, cattle or other animals creates overgrazing. Village common lands, pastures and forestland of the government come handy for the villager's use. Most of the time, it has been found that there are restrictions on the right of entry and land use in these categories, especially on the government owned forestland. The common tendency is towards over-use which destroys the grass and scrubs excessively making the top soil more vulnerable to the monsoon rain leading to the substantial soil loss in the Doon valley. Fodder for these animals also came from the open forest and dense forest through lopping etc. which starts the thinning process of the forest. Livestock census is available for every four years, but we could not get the data for Doon valley beyond 1968 due to weeding out of the old data. Following table provides the details of livestock census.

Table 8.6: Livestock in Doon Valley

Year	*Bovine*	*Buffalo*	*Goat + Sheep*	*Horse-Ponies-Donkeys*
1968	8,3982	34,572	8,5435	1487
1972	11,2982	39,852	10,2822	1892
1978	16,3899	45,553	13,3040	2499
1982	15,9005	51,832	11,7705	2549
1989	16,2484	64,354	11,4151	3959
1993	18,9893	51,527	13,4835	4349
1997	21,1005	48,152	18,0437	4417

Source: Krishi Ganana, Dehradun Tehsil for different years.

All types of livestock have increased significantly over the last 30 years. An attempt was made to collect the information about the livestock as homefed and dependent on the pastureland as these could have provided a better approximation for overgrazing, but this type of classification simply did not exist as was gathered from the Chief Veterinary Officer of the Tehsil. However, the significant growth in total livestock is quite evident and has accelerated the process of overgrazing in the valley.

8.2.6 Fragmentation of Land Holdings

Population explosion has a direct bearing on the size of land holdings. As land holding size gets smaller and smaller, scientific crop management and soil conservation practice becomes more difficult to apply. For Doon valley, land holding size has become smaller over a period of time.

From the table it is evident that percentage of area falling under less than a hectare of total cultivable area is continuously increasing. In 1968, 1.61 per cent of total cultivable areas was of 1 ha or less size, which became 19.86 per cent in 1972, 21.36 per cent in 1978, 27.17 per cent in 1982 and 27.74 per cent in 1986. Area falling under 1 to 3 ha size remains more or less the same around 36 per cent over the period 1968-1986. Farm size of greater than 3 ha area as a percentage of total arable area has continuously declined from 43.4 per cent in 1968 to 35.6 per cent in 1986. This phenomena evidently suggest that the fragmentation of land holding has been taking place in the valley. This process impairs the adoption of modern and mechanical conservation practices by the farmers. And in the absence of proper conservation measures excessive loss of topsoil takes place on continues basis particularly along the hills.

Direct causes of soil erosion emanate from population pressure combined with immigration of people from different parts of the country in the valley. People who are not engaged in industrial activities or services sector fall back on primary activities for their livelihood. This leads to intensive agriculture, continuous and unscientific rotation of crops and operation of cultivation reaches at the steep slope and which are ecologically fragile and highly unsuitable for agriculture. Cultivation on hill having slope greater than 33 per cent has continuously been rising. The period, for which data were available, the area under hill cultivation has more than doubled from 58.39 ha in 1970 to 130.09 ha in 1994. During 1950s/60s this process got started and accelerated. The soil erosion in the cultivated hill sometimes exceeds the figures of over 700 tons/ha/year (*see* Appendix).

Established findings on deforestation causing soil erosion (esp. water-induced) get exemplified for the valley. Over a period of time, dense forest has got converted into open forest and open forest into scrub. This process has led to the process of erosion. In late seventies and eighties some amount of afforestation has blurred this process in the valley but adequate time-series information on plantation/ afforestation is unavailable. Over the period 1970-94 the total forest area has declined from 145973.40 ha to 135829.10 ha.

Population of livestock in the valley has continuously increased. This trend has been continuing since 1960s, the significant rise in the population of goats and sheeps dependent upon pastureland has created the situation of overgrazing causing significant rise in the soil loss.

Soil conservation and land management practices have been unsatisfactory owing to the fragmented pieces of land and in most of the cases the area of those pieces have been less than even an hectare. Declining size of land holding prevent the farmers from adopting sound soil conservation especially along the slopes which in turn causes soil erosion.

Besides these factors, soil erosion in Doon valley has risen over last 25 years due to the absentee cultivation where the cultivable land is owned by the people who do not perform the agriculture but stays away from the land and gives land to the local marginal farmers on contract. Poor local farmer has hardly any incentive to adopt any soil conservation practice. Data related to this fact was absent because in UP this type of system is legally prohibited, but the District Statistical Officer in a candid revelation confirmed that people illegally have been practicing this absenti-landlordism system in Doon valley especially in Rampur, Doiwala and Sahaspur blocks. Absence of strong property rights in cultivation practice accelerates the process of soil erosion in the valley.

8.3 Soil Loss in Doon Valley

A number of parametric models have been developed to predict soil erosion and these models are based on soil, land-use/land cover, landform, and climatic and topographic information. Soil erosion is most frequently assessed by using Universal Soil Loss Equation (USLE) (Wischmeier and Smith, 1965) because this model's major input parameters like soil erodibility, land cover and terrain slope gradient factors can be derived from remote sensing data. A number of studies have demonstrated the capability of satellite data in estimation of soil loss. Spanner *et al.* (1982) demonstrated the potential use of various data sources, *viz.* Landsat MSS, digitised

Table 8.7: Size of Land Holding in Dehradun

Year	*Upto 1 hectare*			*1-3 hectare*			*> 3 hectare*			*Total*	
	No.	*Area*	*Per cent of Total area*	*No.*	*Area*	*Per cent of Total area*	*No.*	*Area*	*Per cent of Total area*	*No.*	*Area*
1968	3080	987	1.61	N.A	N.A	-	N.A	N.A	-	N.A	61008
1972	35192	2186	19.86	12909	22594	36.83	4808	26620	43.40	52909	61340
1978	38120	13240	21.36	12847	22379	36.10	4854	26367	42.54	55821	61986
1982	39272	2630	27.17	9660	16226	34.91	3077	17621	37.91	52009	46477
1986	48852	16414	27.74	14062	21637	36.57	4066	21110	35.68	66980	59161
1990	53619	25501	–	12439	23191	–	3692	18904	–	N.A	N.A

Source: Based on Agricultural Census for 1968, 1972, 1978, 1982, 1986 and 1990.

USDA-SCS soil and NOAA satellite derived precipitation maps and USGS digital terrain data to map various potential soil-erosion hazard classes using USLE.

8.3.1 Methodology

The first objective here is: to prepare an inventory on spatial erosion soil loss in Doon valley using physiographic: soil—land-use map, terrain slope information, rainfall-climatic data and forest cover using USLE. The data on forest cover, land-use, physiography and soils and slope used in this study was provided by CISMHE. The topographic information was interpreted from Survey of India toposheets on 1:50,000 and 1:250,000 scales and rainfall data of last 24 years (1971 to 1994) collected from ground meteorological station.

8.3.1.1 Universal Soil Loss Equation

The estimation of soil loss for Doon valley is based on the Universal Soil Loss Equation (USLE) developed by Wischmeier and Smith (1965). USLE is a quantitative method of predicting soil losses from several natural and anthropocentric factors. The equation is:

$$A = RKLSCP$$

where A = soil loss in tons/ha
R = rainfall erosivity factor
K = soil erodibility factor
L = slope length factor
S = slope gradient factor
C = crop management factor
P = conservation practice factor

For a specific region, A is determined empirically for a range of representative soils on experimental plots by standardizing factors L, S, C and P. L = 1 for a slope 72.5 ft long and S = 1 at 9 per cent gradient. Albeit the equation was developed in the US where experimental plots used with 1/100-acre length, 6 ft width and median slope 9 per cent. C = 1 for bare cultivated fallow and P = 1 for the situations when ploughed up and down the steepest slope. This equation has widely been used in various parts of the world with satisfactory results. Detail discussion of the applications and limitations of the USLE are found in Goldsmith (1977), Hudson (1971), Soil Conservation Society of America (1977), Wischmeier and Smith (1978; In United States Department of Agriculture Handbook No. 537), Elwel and Stocking (1982) and Zachar (1982).

Here, A = RK (the basic equations)

L, S, C and P are dimensionless as they are ratios. The unit of R depends on the means of deriving the index. K is merely a coefficient to balance A and R and its unit is derived from these two.

8.3.1.2 Annual Soil Loss

Annual soil loss provides the estimation of loss of topsoil per unit of area, *i.e.* ha or acre (depending on other units) for every year. It is used for selecting crop management practices by fixing the maximum tolerable limit of soil loss per year. Generally a higher limit of soil loss is presumed for tropical hilly regions where the rate of soil formation is rapid.

The methods used for determining different USLE factors are summarized below.

8.3.1.3 Rainfall Erosivity Factor (R)

It is the localized value of the rainfall-erosion index. Wischmeier found that for cultivated fallow, soil loss was directly proportional to the product of kinetic energy and the maximum 30 minute intensity of the rainstorm. The sum of these products scaled E1 values for a given period provides a numerical rainfall erosion index that evaluates the erosion potential of the rainfall with the given period. Wischmeier concluded that erosion index will be greater for larger drop size of the rainfall and/or velocity of the rain. In practice this has to be calculated from an empirically determined relationship with rainfall intensity. Outside the USA, the correlation of soil erosion with rainfall data has been found to give poor correlation especially in tropical countries. However, this is the most important parameter and the localized empirical data can yield the desired explanation of soil loss. For any region the index used must depend on the rainfall data available and where autographed records exist.

R (Rainfall erosivity factor) value was calculated using following expression given by Bergsma (1980), *i.e.*

$$R = 0.1059 \text{ a.b.c.} + 52$$

where

a = Average annual precipitation (cm);
b = Maximum 24 hrs. precipitation with recurrence interval cf two years (cm/hr); and
c = one hour maximum precipitation with recurrence interval of two years (cm/hr).

The annual R-values, *i.e.* 1066 and 500 for plains and hilly areas respectively were adopted from Saha *et al.* (1992) and Ravishankar *et al.* (1994) as their study areas coincides with the present study area.

8.3.1.4 Soil Erodibility Factor (K)

This factor comes from the inherent quality of soil pertaining to its type, series, texture, etc. This factor in the USLE is tons of soil loss per hectare per unit of rainfall erosion index for a slope of specified dimensions (generally 9 per cent and 22 metre long) under continuous cultivated fallow with any influence of crop cover. Originally K was determined by standardizing L, S, C and P, so that K = A/R. In 1971, Wischemeier *et al.* has produced a nomograph to derive the value for K from the known values.

These values are:

(a) per cent silt and very fine sand (0.002 to 0.1 mm),
(b) per cent sand (0.1 to 2.0 mm),
(c) organic matter content,
(d) top soil structure grade, and
(e) permeability grade.

Following this methodology the value of K for the US, Sri Lanka, West Africa comes out between 0.03 to 0.69, 0.31 and 0.05 – 0.30, respectively.

The K (soil erodibility factor) for each soilscape unit, *i.e.* physiography—soil and land-use association units was calculated using the data from field surveys as well as analysis of soil samples in laboratory. Physical-chemical parameters, *viz.* soil texture, soil structure, organic matter content and permeability of surface soil samples of each soilscape unit were derived following the nomograms given by Wischmeier and Smith (1965).

8.3.1.5 Length of Slope Factor (L)

It is the ratio of soil loss from a specific length of slope to that from the length specified for the K factor of the equation. Slope length being defined as the distance from the point of origin of overland flow to either of the following, whichever is limiting for the major portion of the area under consideration

(a) the point where the slope decreases to the extent that deposition begins, or
(b) the point where run-off enters a well-defined channel which may be part of a drainage network or a constructed channel such as terrace or a diversion.

8.3.1.6 Steepness of Slope Factor (S)

This factor is the ratio of soil loss from a specific slope per cent to that from the slope per cent specified for the K value of the equation. The effect of slope factor (combined with L) is more pronounced in the tropical countries like India and its value derived in temperate countries should not be used indiscriminately.

Derivation of LS (slope length and slope factor) value for each soilscape unit was performed by computing the slope length and gradient using Survey of India toposheets at 1: 50,000 scale, checked by field measurements wherever necessary and by following the nomograms of Wischmeier and Smith (1965).

8.3.1.7 Crop Management Factor (C)

This factor combines the effects of crops, crop sequence and the various management practices. It is defined as the ratio of soil loss from land cropped under specific conditions to the corresponding loss from bore tilled fallow. It reflects the protective influence of the vegetation and ground cover. It is possibly the most complicated factor in the equation. It can either be derived directly from the trial plots by conducting experiments or from the Wischmeier-Smith procedure. They divide the cropping year into five periods:

Period F } (rough fallow)
Period 1 } Establishment of crop
Period 2 }
Period 3 } Growing and maturing crop
Period 4 } Residue or stubble

For each period a value, c, is derived and multiplied by the relevant proportion of the annual rainfall erosivity, r, from a cumulative distribution curve (Hudson, 1971).

$$\text{Thus, } C \times R = c_f r_f + c_1 r_1 + c_2 r_2 + c_3 r_3 + c_4 r_4$$

This method is more sensitive than simply using annual c and r-values as both vary in an interrelated way throughout the year. For instance a season of high erosivity may produce little erosion if it coincides with a mature crop with good ground cover. The cover factor (C) value for each soilscape unit was determined using information on land-use/land-cover provided by CISMHE following the tables given by Wischmeier and Smith (1965).

8.3.1.8 Conservation Practice Factor (P)

This is the ratio by which erosion is reduced by conservation measures from the worst possible case (P = 1.0), in which soil is ploughed up and down the steepest slope. This factor is closely related to the slope factor. The conservation factor (P) values were chosen based on field survey information provided by CISMHE, research findings of Central Soil and Water Conservation Research and Training Institute, Dehradun and from studies of Saha *et al.* (1992) and Ravishankar *et al.* (1994).

8.3.2 Data Used in Estimation of Soil Loss in Doon Valley

8.3.2.1 Data Used

For evolving the data on land-use category as far as satellite data of

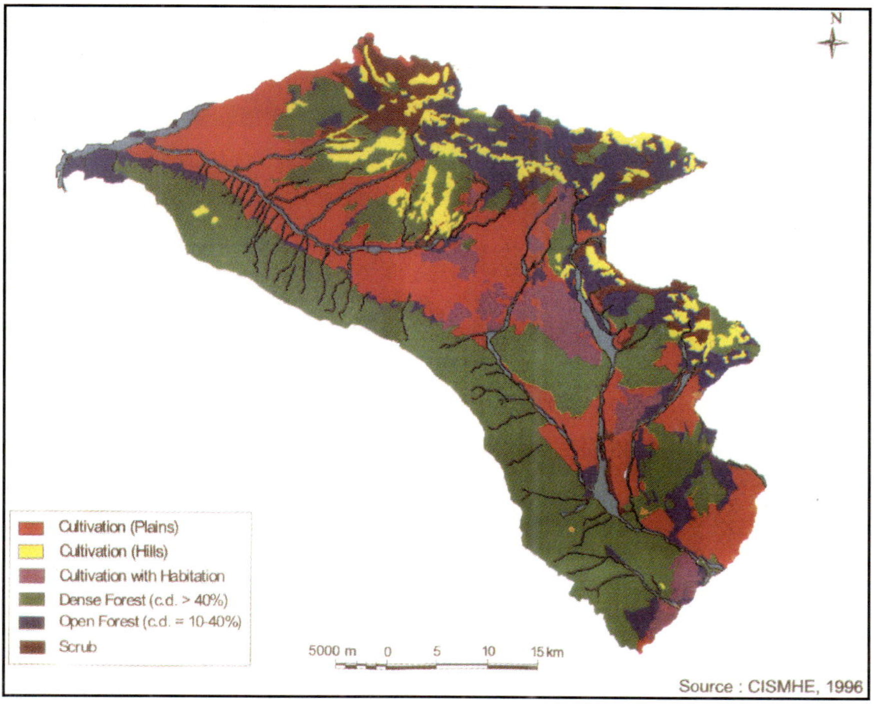

Fig. 8.2: Landuse/Land Cover Map of Doon Valley

the years 1972, 1974, 1977, 1979, 1982, 1993 and 1994 was used. The interpretation of data was performed for the following land-use categories.

(a) Dense forest (crown density > 40 per cent)
(b) Open forest (crown density = 10-40 per cent)
(c) Scrub/Grassland (crown density < 10 per cent)
(d) Cultivation in hills
(e) Cultivation in plains
(f) Cultivation with habitation

Land-use categories were extracted from Landsat FCCs of December 1972, April 1979 and March 1993 and ERTS imagery (B&W) of May 1974 and 1977 on 1:250,000 scale and IRS FCC of January 1994 on 1:50,000 scale were used. The interpreted information from satellite data and GIS analysis was obtained from the Centre for Inter-disciplinary Studies of Mountain & Hill Environment (CISMHE), University of Delhi, South Campus.

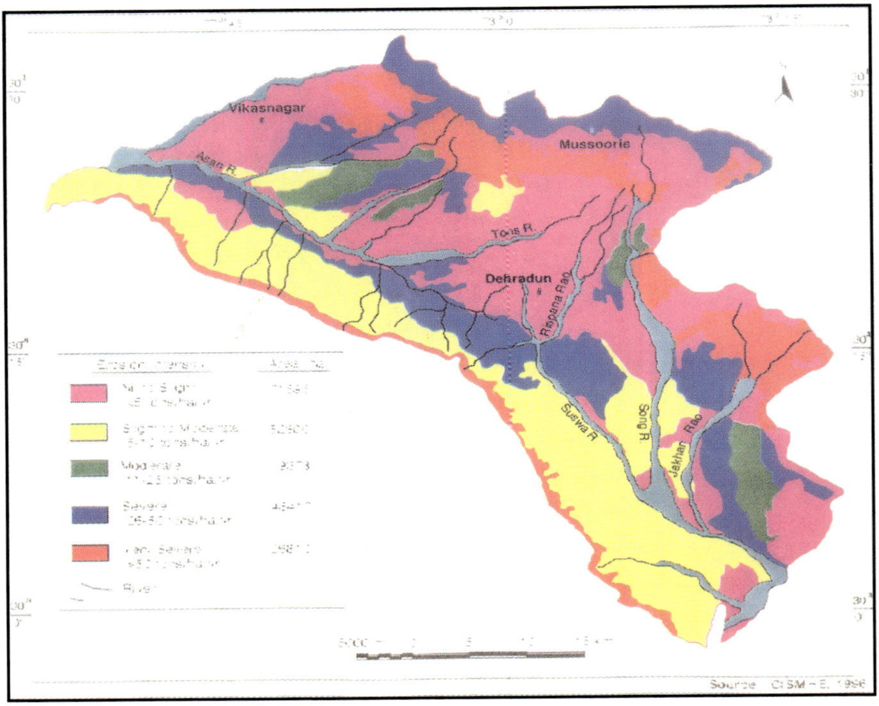

Fig. 8.3: Soil Erosion Hazard Classes Index Map of Doon Valley

8.3.2.2 Data Constraints

For estimation of soil loss in Doon valley region, first of all data on land-use cover was developed. Data on land utilization statistics are generally provided by the office of District Statistical Officer through a bulletin known as 'Sankhiyaki Patrika'. But contrary to general use of this source, we have their classification of land utilization, which is not concomitant to the requirement of USLE. It is also highly unscientific and its computation procedure is full of lacuna as explained earlier. Due to this, data were generated from satellite imageries. For different years the scales were different and this caused some amount of subjectivity in the interpretation of the imageries. For example, on a smaller scale, it was very difficult to distinguish thin forest from plantation. However, utmost precautions were taken and occasionally expert's opinions were also incorporated.

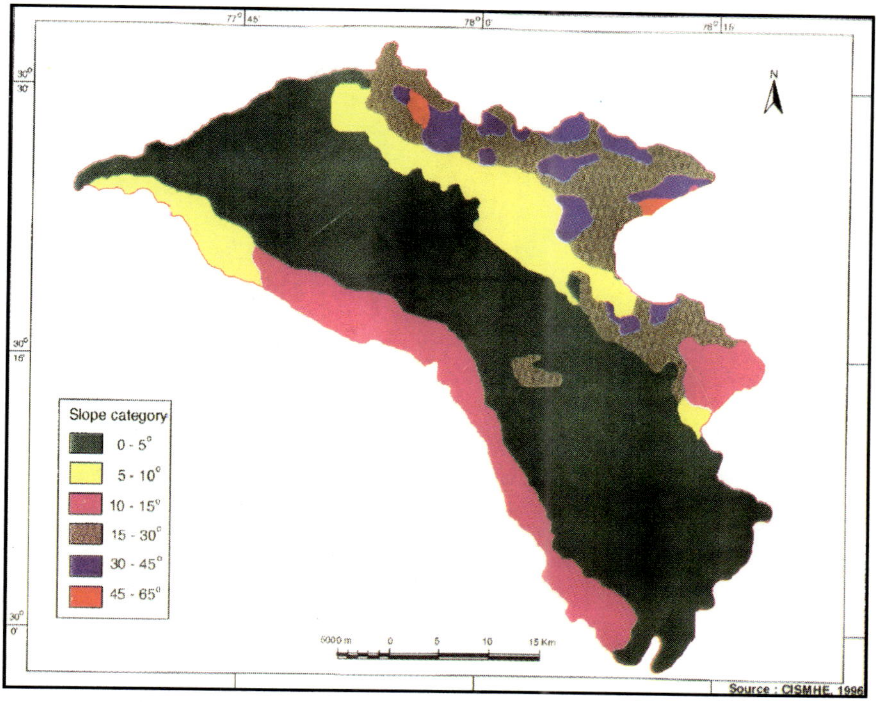

Fig. 8.4: Slope Category Map of Doon Valley

8.3.2.3 Land-use

As C and P are the most important factors of USLE, therefore, land-use for different years was calculated. The interpretation of IRS data on 1:50,000 and Landsat MSS data on 1:250,000 scales using yielded the following data on land-use category for years 1972 and 1992 (see Table 8.8).

Table 8.8: Land-use Pattern in Doon Valley in 1972 and 1992

Land-use	Area (ha)	
	1972	*1992*
Dense forest	90444	84390
Open forest	20924	35458
Scrub/Grassland	24171	25717
Cultivation (Hills)	6558	13034
Cultivation (Plains)	60636	45354
Cultivation with habitation	6271	5048
Total	**209004**	**209001**

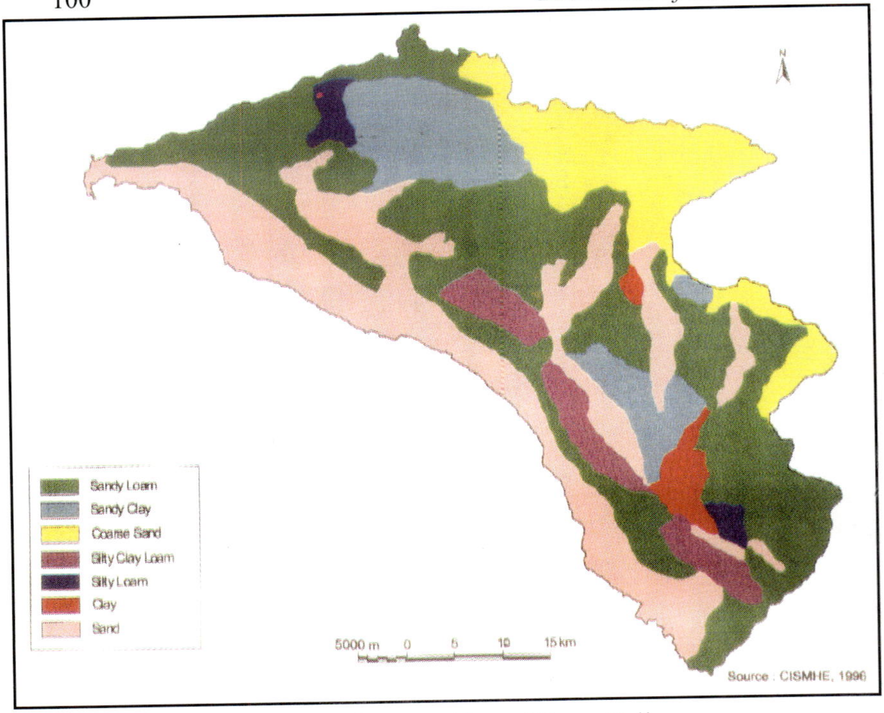

Fig. 8.5: Soil Types Map of Doon Valley

The total area of Doon valley stands around 209000 ha. marginal fluctuations in the area can be attributed to the error arising out of visual interpretation and difference in the scale of imageries.

8.3.3 Earlier Studies on Soil Loss in Doon Valley

In the first study by Saha et al. (1992), primarily part of Song river watershed comprising eastern part of Doon valley has been considered. The types of data used in the study are: CCT of IRS-1A-LISS II of Path No. 28 Row No. 46 dated May, 1988 and FCC of same on 1:250,000 scale; Survey of India topographical maps on 1:50,000 and 1:250,000 scales and rainfall data of last 20 years, i.e. 1971-1990 collected from ground meteorological station. Further, digital analysis was carried out in the study using VIPS-32 image processing software package on VAX 11/780 computer system at RRSSC (Regional Remote Sensing Service Centre), Dehradun. This study also prepared a digital colour coded physiography-soil-landuse map by following the supervised classification approach using maximum likelihood classifier, supplemented with field survey information on morphogenesis of soils in various soilscape units. Subsequently the value of USLE parameters was calculated.

For the Song river watershed, the annual values of R were determined as 1066 for plains and 500 for hilly areas. For estimating the K factor, physiography—soil and land-use association for each soilscale unit was calculated using field and laboratory estimated soil physico-chemical properties data, *viz.*, soil texture, soil structure, organic matter content and permeability. For this purpose the monopgraph prepared by Wischmeiere & Smith (1965) was extensively used in this study. Deriving the value for each soilscale by computing slope length and gradient using SOI topographical maps on 1 did estimation of LS factor: 50,000 scale. It was further checked by field measurement following the norms of nomogram of Wischmeier and Smith (1965). The P factor was derived from the field observation and research findings of Central Soil and Water Conservation Research and Training Institute, Dehradun. The C factor was estimated from information based on digital soil land-use map.

In another significant study by Ravishankar *e. al.* (1994), Sitla Rao watershed area constituting western part of Doon valley was considered. The data used for this purpose were of various types but it comprised mainly a CCT of IRS-IB LISS-I data of path 29 row 46 dated 24 October, 1992; an FCC of TM data on 1:50,000 scale (19 April, 1990; path 146, row 39) and a Survey of India (1:50,000 scale, 53 F/15) topographic map. Digital analysis was done on a similar system used in the earlier studies at RRSSC, Dehradun.

This study calculated the parameters of USLE by following the same methodology as followed by Saha *et al.* (1992). The value for R, K and LS were more or less the same as obtained by the Saha's *et al.* study.

8.3.4 Estimation of Soil Loss in Doon Valley

The detailed results of digital physiography, soils and land-use analysis of Doon valley are given in Appendix I. The soilscape units as delineated and interpreted from IRS IB LISS-II data have been described below and classified digital physiography-soil map is given in Fig. 8.4.

The erosional soil loss data thus calculated for entire Doon valley was integrated into digital soil-physiography map and the resultant map thus obtained was colour coded to generate a soil erosion assessment map with different soil erosion intensity classes (Fig. 8.5).

The soil erosion intensity formed the basis of devising appropriate conservation measures to be adopted in the valley and have discussed later under the heading 'Conservation Measures'.

All the physiographic USLE parameters, *viz.* R (Rainfall erosivity factor), K (Soil erosivity factor) and LS (Slope length and slope factor) remain constant for Doon valley, only the parameters like C (Cover factor) and P (Conservation factor) change during the course of time. Therefore, from the soil loss estimates generated as described above, year-wise total annual soil loss in the entire Doon valley was calculated by multiplying rate of annual soil loss in tons per hectare with area under a particular land-use category in different years, *i.e.* in the years 1972, 1974, 1977, 1979, 1982, 1993 and 1994. For this purpose satellite data was used and interpreted to generate required land-use classes, *viz.* Dense forest (Crown density > 40 per cent), Open forest (c.d. = 10-40 per cent), Scrub/Grassland (c.d. <10 per cent), Cultivation in hills, Cultivation in plains, Cultivation with habitation and Mined areas wherever possible. For the analysis of land-use Landsat FCCs of December 1972, April 1979 and March 1993, and ERTS Imagery (B&W) of May 1974 and 1977 on 1:250,000 scale and IRS FCC of January 1994 on 1:50,000 scale were used. The interpreted information from satellite data and GIS analysis was obtained from CISMHE.

The values for rate of soil loss for different years were derived using the data on soil loss given at Appendix 1. It gives the summary of land-use type without providing the slope class and 'aspects' along which the value of USLE parameters vary. For this purpose, Geographical Information System (GIS) based exercises were conducted at CISMHE, University of Delhi South Campus where different maps of slope, aspects types of forests (sal etc.), conservation factor practices were superimposed on the land-use category (Fig. 8.6). From this superimposition, area under dense forest area lying on different slopes and different aspects. From the Appendix I it is clear that dense forest on hills (slope class: 30-50 per cent) the value for USLE parameters, *i.e.* R, K, LS, C and P are 500, 0.38, 7.1, 0.005 and 0.5 respectively. From these values the annual soil loss (A) comes out as 3.37 tons/ha/year.

$$\text{Since } A = R\,K\,LS\,C\,P = 500 \times 0.38 \times 7.1 \times 0.005 \times 0.5 = 3.37$$

But the same dense forest on the same slope, *i.e.* 30-50 per cent but on the mountains gives the different value due to the difference in erodibility factor (*K*) and slope factor (*LS*) here the value of annual soil loss comes to be 2.93 t/ha/year.

$$\text{Since } A = R\,K\,LS\,C\,P = 500 \times 0.39 \times 6.00 \times 0.005 \times 0.5 = 2.93$$

Further the dense forest along the upper piedmont (4-7 per cent slope) gives the value of annual soil loss around 0.7 t/ha/year due to the variation in USLE parameters. Similarly it can be demonstrated that for the same land-use category variation in physiography and slope the parameters of USLE will vary and it will induce further variation in the figure for annual soil loss.

With the help of soil scientists of Central Soil and Water Conservation Research and Training Institute, Dehradun and CISMHE, University of Delhi, using GIS, the average (weighted average) of soil loss/ha/year was calculated for each land-use category for the different years. The computation of soil loss for each category was done at 13 observation points starting at 1970 and ending at 1994. The range of annual soil loss for each category has been summarized in Table 8.9.

After obtaining the figure for annual soil loss for period 1970, 1972, 1974, 1976, 1978, 1980, 1982, 1984, 1986, 1988, 1990, 1992 and 1994 the corresponding area has been multiplied by this figure. For instance year 1972 and 1992, Table 8.10 provides the figure for total soil loss in the Doon valley.

Table 8.9: Annual Soil Loss (T/Ha) for Each Land-use Category (See Appendix I)

Land-use Category	*Range of Soil Loss (t/ha/yr)*	
	Minimum	*Maximum*
Dense forest	3.52	3.45
Open forest	34.72	34.06
Scrub/Grassland	623.58	611.23
Cultivation Hills	767.54	611.69
Cultivation Plains	9.33	9.15
Habitation	1.11	1.08

Table 8.10: Soil Loss in Doon Valley in Year 1972 and 1992.

Year	*Landuse Type*	*Area (ha)*	*Rate of soil loss (t/ha)*	*Total Loss (tons)*
		A	B	C
1972	Dense forest	90444	3.51	316624.00
	Open forest	20924	34.61	724179.64
	Scrub/Grassland	24171	621.00	15010191.00
	Cultivation (Hills)	6544	766.04	5012935.20
	Cultivation (Plains)	60636	9.30	563914.80
	Habitation	6281	1.10	6909.10
	Total			**21634753.74**
		~21.634 million tons		
1992	Dense forest	84390	3.49	294794.73
	Open forest	35458	34.54	1224848.90
	Scrub/Grassland	25717	619.79	15939207.00
	Cultivation (Hills)	13034	762.93	9944079.60
	Cultivation (Plains)	45354	9.28	420983.10
	Habitation	5058	1.10	5540.70
	Total			**27829452.80**
			~27.829 million tons	

Similarly, computation of total soil loss has been done for the Doon valley for different years and has been given in Table 8.11.

Table 8.11: Temporal Soil Loss in Doon Valley

Year	*Total Soil Loss (in million tons)*	*Soil Loss (tons/ha)*
1970	21.454	102.65
1972	21.634 (0.839)	103.51
1974	21.769 (0.624)	104.16
1976	24.421 (12.182)	116.85
1978	23.515 (–3.709)	112.51
1980	23.848 (1.416)	114.10
1982	23.885 (0.155)	114.28
1984	23.900 (0.062)	114.35
1986	23.909 (0.037)	114.40
1988	24.001 (0.384)	114.84
1990	24.207 (0.858)	115.82
1992	27.829 (14.962)	133.15
1994	27.909 (0.287)	133.54

Note: Figures in parenthesis shows the percentage change over the previous period.

Soil erosion in the area of Doon valley is showing a continuously increasing trend. From the table it is eminently evident that except the year 1978, the erosion has registered an average growth rate of 0.258 per cent per annum. In 1970 the loss of top soil was 21.454 mill/tons which reached 27.909 tones in 1994 showing a percentage increase of around 30 per cent over 25 years period of time. On per ha basis in 1970 the loss of topsoil was 102.65 tons, which reached 116.85 tons/ha in 1976, and in 1994 it became 133.54 tons/ha. On an area of valley showing the slope of even 100 per cent and possessing the characteristics of alluvial soil, a maximum of 5-10 tons/ha can be considered as tolerable limit but the present rate of erosion exceeding 133 t/ha/yr. is much higher than the tolerance limit threatening the very basis of agriculture.

Taking into account of the crop management factor and conservation factor the estimation of topsoil loss in the valley has been done. Since other factors are physiographic in nature they have been presumed to be constant.

On the basis of the parametric of USLE, it has been found that the maximum erosion takes place on the area of hill cultivation. Appendix I provides the range of 767.54-611.69 t/ha/yr. For cultivation hill area this is maximum and has continuously been increasing in the valley. In 1970 it was around 5,458 ha which grew to 13,009 ha in 1994. In 1994 it accounted for more than 35 per cent of the total soil loss in the valley. Scrub or grassland is also experiencing the erosion in the range of 623.58-611.23 t/ha/yr. The area under this also has registered a continuous growth over 1970-1994 periods. Dense forest has declined but not significantly. Moreover for dense forest cover the loss is around 3.51 t/ha/yr, which is much within the tolerance limit. Area of dense forest could have declined much further had the plantation not taken place in the valley in 1980s and 1990s. Overall, thinning of forest area and encroachment of fragile hill for cultivation purposes are causing the maximum damage to the erosion in the valley.

The values for rate of soil loss for different years were derived using the data obtained in Appendix I. As the values of rate of soil loss given in the Appendix I for different land-use categories varies from one slope class to another slope class and also the aspect of a particular slope. The rate of soil loss values for a particular land-use category for different years were calculated with the help of GIS analysis using land-use, slope maps and soil erosion class maps of

Doon valley. Annual soil loss thus calculated is given in Table 8.11. Land-use map of Doon valley of the year 1994 has been given at Fig. 8.4. The map of Doon valley with soil erosion classes is given in Fig. 8.5.

The data pertaining to USLE factor values, soil loss estimates and soil erosion intensity classes for each soilscape unit are presented in Appendix I. It is observed that most erodible soils (K range from 0.39 to 0.5) predominate in terraces, floodplains and Steep Mountain slopes. The data also indicates that steep to very Steep Mountain soilscape units have higher LS values while other soilscape units due to their lesser slope (S) and slope length (LS) have lower LS values ranging from 0.1 to 1.0. The conservation factor data indicate that scrub and low intensity cultivation areas in the valley are prone to higher soil losses. It is the densely forested areas only which are least prone to soil erosion. Therefore, steep to moderately sloping, barren mountain slopes or under scrub cover and hill slopes under terrace cultivation contributes to severe to very severe soil erosion in the valley.

9

Economic Estimates of Soil Erosion in Doon Valley

Economic valuation of environmental damage is needed not only to correct the market failure but also to change the behaviour of economic agents at micro and macro levels. Economic estimates of lost soil is necessary for determining the allocation of resources needed to analyse and remedy the problem of soil erosion. It is also required to arrive at the better approximation and estimates of National Income Accounts at the national level.

9.1 Methodology and Empirical Evidences from Developing Countries

In large number of developing countries of Africa and Asia, valuation of eroded soil has been done following different measures of economic losses. The ways, in which the cost of erosion has been measured, generally vary across the regions and they need some detailed and close scrutiny. A review of these studies would revolve around the following issues:

9.1.1 Gross Vs. *Net Cost*

Most of the studies available on the economic cost of soil erosion are 'gross' in nature because neither the costs are directly compared to investment costs for conservation that would mitigate the losses due to soil erosion nor soil erosion has been accounted. Studies like Bishop and Allen (1989), Bojo (1991) and Grohs (1994) do make adjustments in variable costs, *i.e.* harvest, weeding, transport etc. due to falling yields but these studies hardly say anything about the change in fertilization, crops growth etc. for which data are difficult to obtain.

9.1.2 Short-run Vs. Long-run Costs

The process of erosion removes the top soil in varying degree and with it goes the nutrients, water holding capacity and other organic matter. Some studies put economic values for eroded soil at a particular point of time. But erosion of top soil affects the growth of plants and crops in future for many years even if the soil depth is not in the critical range. The loss of this natural capital reduces the future income stream for the farmers for many years to come and hence longer time horizon is required in place of one time loss of productivity due to the erosion.

9.1.3 On-site Vs. Off-site Costs

By and large every available studies in developing countries on the economic valuation considers the on-site effect of soil erosion. Impacts of soil erosion like downstream siltation etc., which come under off-site category, have hardly been considered primarily due to paucity of adequate database. A study like of Zimbabve (Groh, 1994) is exception where off-site impacts have been considered in a significant manner.

9.1.4 Other Features

Most of the studies on valuation of soil erosion primarily considers water-induced erosion only. Few studies, *e.g.* South India (Barandon & Hommann, 1995), Africa (McKenzie, 1994), go beyond the boundary of erosion and consider all types of land degradation, *i.e.* waterlogging, soil salinity etc. Values of the estimates made for different countries have been expressed in different terms. Most of the values have been expressed either as a percentage of agricultural gross domestic product or simply gross domestic product. Following is the table showing the magnitude of economic losses due to land degradation (other form of degradation has been integrated with erosion in valuation process).

Studies in the table follow different approaches to valuation and differences in approach sometimes become the basis of variation in the estimates made.

Table 9.1: Economic Estimates of Land Degradation

Country	*US $ mill. At 1994 price*	*Per cent AGDP·*	*Source/Study*
Zimbabwe	117	9	Stocking, 1986
Ethiopia	115	5	Sutcliff, 1993
Malawi	6.6-19.0	3	World Bank, 1992
India	1500-2400*	4-6.3	World Bank, 1995
South Africa	18	<1	McKenzie, 1994

* At 1994 price, others at current price.

Source: Author's compilation.

9.2 Approaches to Valuation of Soil Erosion

For calculation of cost of soil erosion to individual farmers or society, several methods have been proposed in various studies. Most of the cost estimation relies upon the replacement cost of nutrient lost. A few studies are based on Hicksian notion of income (Hicks, 1946) where all soil lost are not necessarily a cost inflicted on individual or society. The reason forwarded is that if technological development takes place in the economy, erosion of soil can be allowed without compromising the future consumption. Under this approach, soil is considered as a form of wealth providing an income stream and hence cost of soil erosion is calculated as the present value of future income loss due to soil erosion. However, in the current study replacement cost (resource value) approach has been adopted because for income (production value) approach, the required experimental data simply does not exist.

Soil can be seen as a source of inputs for agricultural output or it can be seen as a productivity-deciding factor. First consideration gives rise to the methodology where loss of input is replaced by artificial compensating inputs having well established market price and is known as replacement cost approach. Second consideration tries to measure the effect on productivity of the agricultural crops again at the conventional market price and it is known as productivity loss approach.

9.2.1 Replacement Cost Approach

Under this approach first of all average rate of soil loss per geographical area is calculated, generally by using Universal Soil Loss Equation (USLE), then corresponding losses of nutrients are calculated by using regressions developed on the basis of experimental data. Thereafter this loss of nutrients per hectare is evaluated at the market price at which the replaceable commercial fertilizer is purchased. This process can be used to arrive at the national figure of the value of soil loss.

It has been found that only major nutrients like Nitrogen (N), Potassium (K) and Phosphorous (P) are taken into account. Minor nutrients like Magnesium, Calcium, Organic material etc. are ignored. But soil is not just a provider of these nutrients; it is a complete ecosystem in itself. In this kind of valuation, other important ecological functions of soil are ignored. The data requirement under this approach is mammoth. Stocking (1986) calculated the economic cost of eroded soil for Zimbabwe in terms of three major nutrients *i.e.* N, P and K.

9.2.2 Productivity Loss Approach

The complex character of soil and interplay of large number of natural factors like rain, climate, altitude etc. impairs the relationship of soil erosion and productivity of crops with precision and accuracy of satisfactory order. Various methods under productivity loss approach can be adopted.

9.2.2.1 Inferred Soil Loss Yield Decline Functions

Under this, it is presumed that erosion reduces productivity through loss of plant-available soil water capacity. Most of the time the exact relationship is derived from the local experimentation centre of the country concerned. A general expression has been derived and widely accepted for this purpose comes from Lal's (1981) experiments as follows

$$Y = ce^{-BT}$$

where

Y	=	Yield in tone per hectare
c	=	Yield on uneroded soil
e	=	Natural log
T	=	Cumulative loss of soil per hectare
B	=	Decline coefficient

9.2.2.2 Soil Depth Loss Yield Decline Models

This method was evolved for Ethiopia (Sutichiff, 1993) and it seems to be improved upon others and involves the following steps:

(i) Calculation of erosion rates by using modified USLE.
(ii) Minimum soil depth beyond which crop production is impossible and the maximum required depth beyond which erosion does not effect crop cultivation is calculated.
(iii) A Water Requirements Satisfaction Index (WRSI) is developed relating monthly values of rainfall, water holding capacity, evapotranspiration etc. The value of the WRSI is mapped to relative reductions in crop yield.

9.2.2.3 Plant Growth Models

Under this model of valuation of soil loss the distribution of sheet erosion is sketched and following the Erosion Productivity Impact Calculator (EPIC) impact of a whole set of physical and chemical characteristics of the soil is observed. By using CERES model, impact of reduced soil rooting depth on plant yields, maintaining all factors as constant is calculated. Finally, income losses for farmers are estimated using crop budgets for all major crops.

Sometimes, in case of unavailability of satisfactory data experts' opinion on erosion and productivity is invited and cost of erosion following productivity approach is calculated.

9.2.3 Expressing the Measurement of Loss

Calculation of economic loss arising out of soil erosion can be expressed in many ways. Prominent among them are:

9.2.3.1 Gross Annual Immediate Loss (GAIL)

This is the annual value of output lost during a year due to the erosion in the previous year.

GAIL = P.dQ
P = Economic price per ton
dQ = Current production in tons loss due to soil erosion in the previous year

Sometimes net instead of gross is used and Net Annual Intermediate Loss (NAIL) is estimated.

NAIL = (P – MC) dQ
MC = Marginal cost per tons saved due to lower yield

9.2.3.2 Gross Discounted Future Loss (GDFL)

Soil erosion is irreversible for the relevant time horizon and the loss in a particular year affects the production for many years.

$$\text{Here, } GDFL = GAIL \; \frac{(1+r)^{n-1}}{r\,(1+r)^{n}}$$

n = The number of years considered
r = The real social rate of discount

If the time horizon (n) is infinite then

$$GDFL = GAIL \; \frac{1}{r}$$

9.2.3.3 Gross Discounted Cumulative Loss (GDCL)

Soil erosion is a cumulative process. Here each year's erosion is followed by another adding layers of cost on top of each other.

$$GDFL = \frac{PdQt}{t = 1\,(1+r)^{t}}$$

and

$$dQ_t = KtQ_b$$

K = A relative decline factor (assumed constant)
t = Time index in years
Q_b = The crop production in the base year

9.3 Valuation Methodology for Soil Erosion in Doon Valley

For valuation of lost soil in the valley area, both the approaches have been adopted. This will enable us to make a meaningful comparison of the results under two approaches. The approaches in consideration are:

9.3.1 Resource Value of Soil

Soil is considered as supplier of useful nutrients. Artificial fertilizers are used as a marketed proxy for valuation of nutrients lost due to erosion. However, resource value of soil differs from the replacement cost approach. Under replacement cost approach of soil erosion, the depletion of nutrients are valued in terms of annual marginal cost of their replacement. The resource value is based on the cumulative depletion of soil nutrient content caused by erosion. It is cumulative to account for erosion progressively impairing the quality of the soil over time and imposing limitations on the soil's potential uses. The depletion of nutrients of soil can be approximated by a negative exponential function and has further been approximated as straight-line soil.

Let us assume that s is the erosion rate, Cu is the soil's content of nutrient u and m is the sediment enrichment ratio, then total losses of nutrients u occurred in year n

$$D_{un} = s\, c_u\, m_n$$

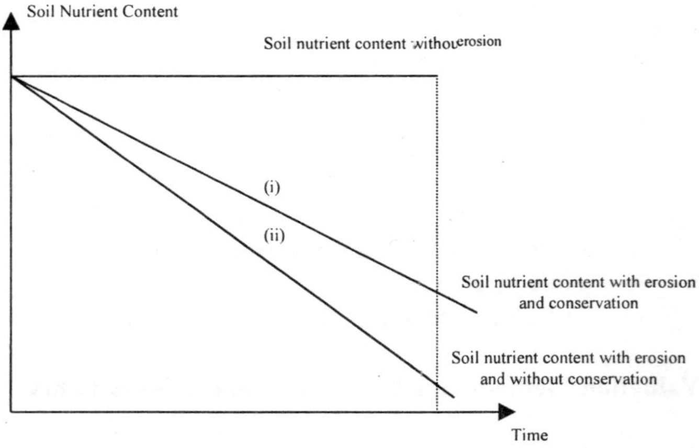

Cumulative soil nutrient loss due to erosion = (i) + (ii)
Cumulative loss of soil nutrients prevented by Conservation = (ii)

Fig. 9.1: Model of the Impacts of Erosion and Conservation on Nutrient

All nutrients let us say they are u, v and w loss due to the erosion in year n

$$A_n = D_{un} + D_{vn} + D_{wn}$$

Over the year 0 to N, total cumulative depletion of the sol's nutrient content

$$A_T = \sum_{n=0}^{N} (D_{um} + D_{vn} + D_{wn})$$

If P_u, P_v and P_w be the market price (after adjusting the subsidies) of the compensative equivalent fertilizers u, v and w, r be the real social rate of discount, then the resource value of soil erosion (B_T) over period 0 to N years.

$$B_T = \sum_{n=0}^{N} \frac{(P_u D_{un} + P_v D_{vn} + P_w D_{wn})}{(1+r)^n}$$

9.3.2 The Production Value of Soil Erosion

It is assumed that under other factors (climate, pests, diseases, fertilisers rates, crop management, technology, etc.) remaining the same, yield of crops gets reduced due to erosion. Further it is also assumed that farmers do not change their cropping pattern due to soil erosion. The relationship between soil erosion and decline in crops yield can be approximated by a negative exponential function (Stocking and Peake, 1986). If Y_n be the yield in year n, Y_o the yield in the base year and d is the annual rate of yield decline then,

$$Y_n = Y_o (1 - d)^n$$

In Fig. 9.2 the relationship between yield and time has been shown through a linear function

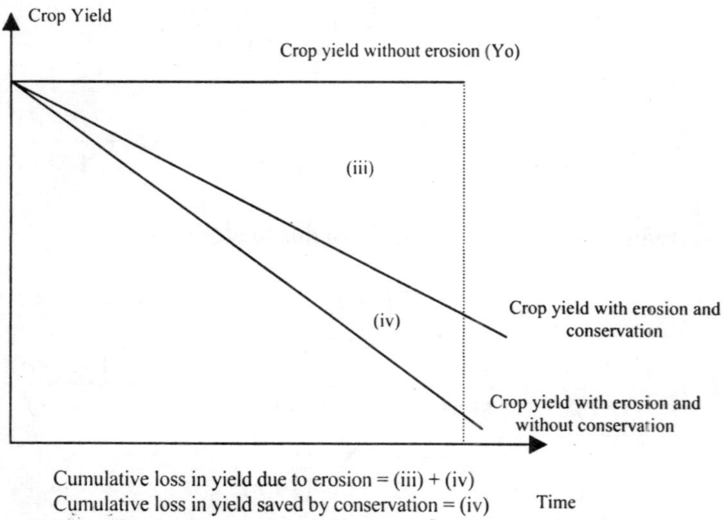

Fig. 9.2: Model of Impacts of Erosion and Conservation on Crop Yield.

if r be the real rate of discount then without any erosion taking place and affecting the yield, the present value of total yield (T) at producer's price P_c, over n years to time horizon.

$$T = \frac{P_c Y_0}{(1+r)} + \frac{P_c Y_1}{(1+r)^1} + \frac{P_c Y_2}{(1+r)^2} + \ldots\ldots + \frac{P_c Y_n}{(1+r)^n}$$

But in case of no erosion yield is supposed to be constant presuming all other natural and man-made factors as unchanging.

Therefore, $Y_0 = Y_1 = Y_2 = \ldots. = Y_n$

$$\Rightarrow T = P_c Y_0 \frac{1}{(1+r)^0} + \frac{1}{(1+r)^1} + \ldots\ldots + \frac{1}{(1+r)^n}$$

$$T\hat{e} = \frac{P_c Y_0}{(1+r)^0} + \frac{P_c Y_0(1-d)}{(1+r)^1} + \frac{P_c Y_0(1-d)^2}{(1+r)^2}$$

$$+ \ldots\ldots\ldots + \frac{P_c Y_0(1-d)^3}{(1+r)^3} + \frac{P_c Y_0(1-d)^n}{(1+r)^n}$$

$$= P_c\ Y_0\ \frac{1}{(1+r)^0} + \frac{(1-d)}{(1+r)} + \frac{(1-d)^2}{(1+r)^2} + \ldots\ldots + \frac{(1-d)^n}{(1+r)^n}$$

therefore present value of yield lost due to erosion

$$T\hat{n} = T - T_e$$

$$= P_c\ Y_0\ \frac{1-(1-d)}{(1+r)^0} + \frac{1-(1-d)^2}{(1+r)^2} + \frac{1-(1-d)^3}{(1+r)^3} + \ldots\ldots + \frac{1-(1-d)^n}{(1+r)^n}$$

$$= P_c\ Y_0\ \sum_{t=1}^{N} \frac{(1-d)^t}{(1+r)^t}$$

$$= \sum_{t=1}^{N} P_c\ Y_0\ \frac{(1-d)^t}{(1+r)^t}$$

present value of yield lost due to erosion.

Now, the present value of yield lost due to soil erosion over N period can be expressed as

$$J_T = \sum_{n=0}^{N} \frac{NP_c\,(Y_0 - Y_0(1-d)^n}{(1+r)^n} \qquad \text{(I)}$$

In the expression (I) values for Y_0, d, r and n will be substituted and the value of loss be obtained.

The value of Y_0 and r has been derived on the basis of information and database of District Statistical Officer, Dehradun and State Planning Institute, UP. The value of d has been derived on the basis of experimental data of National Bureau of Soil Survey and Land Use Planning, Nagpur and Central Soil and Water Conservation Research and Training Institute, Dehradun. Details of the modus operandi of derivation of the values for the variables in the expression has been given in the next section.

9.3.2.1 Data and Discussion

For estimation of total cumulative loss in yields of crops in Doon valley following variables are to be known:

9.3.2.1.1 Yield in the Base Year (Y_o)

The base year has been taken for year 1994 since it is the reference period for all purposes in the study. The yield of wheat, rice and maize has been taken into account, which is the dominant crop in the valley. These three crops occupy around 79 per cent of the total area under cultivation (Sankhikia Patrika, Dehradun, 1996). The yield for three years crops for 1994-95 have been given below.

Yield of Major Crops (Quintal/ha)

Crops/Year	*1994-95*
Wheat	15.74
Rice	17.00
Maize	11.17

Source: Sankhikiya Patrika, Dehradun, UP Planning Unit.

The yield for rice, wheat and maize are not free from erosion affect neither are insensitive to farmers altered response and practice due to erosion in the valley still the average yield serves the purpose.

9.3.2.1.2 Rate of Yield Decline (d)

It shows the rate of decline in the yield of different crops due to the erosion of topsoil, which exposes the subsoil level having lesser nutrients, organic matter and water holding capacity. Experiments conducted by soil scientists in the valley demonstrate that for every cm of topsoil loss, the yield declines and it is maximum for wheat followed by maize, wheat and mustard (Khybri *et al.*, 1980). In another experiment (Agnihotri *et al.*, 1998) it has been found that soil loss causes yield decline maximum for wheat by 16.6 kg ha^{-1} for every cm soil removed. In a more general term, National Bureau of Soil Survey and Land Use Planning, Nagpur (1990) provides figures for decline in yield for different erosion category in different soil type.

Table 9.2: Loss of Yield (Per Cent) of Major Soil Groups Under Different Degree of Erosion

Degree of Erosion	*Annual Erosion (tonnes/ha)*	*Loss (per cent) in yield*		
		Alluvial soil	*Black soil*	*Red soil*
Slight	0-10	0-5	0-10	0-25
Moderate	11-20	5-10	11-25	26-50
Strong	21-40	11-25	26-50	> 50
Severe	> 40	26-50	> 50	N.A

Source: National Bureau of Soil Survey and Land Use Planning, Nagpur, 1990.

By and large Dehradun valley can be characterised as having alluvial soil where average erosion for the whole valley region is around 133.54 t/ha/year, which is much higher than the severe erosion category. For this category of erosion the rate of yield decline for wheat has been found as 40 per cent, for rice 34 per cent and for maize 36 per cent per annum (NBSSLUP, 1990). This is the most conservative estimate and the figure can go much higher than this depending upon some other natural factor like rainfall etc.

9.3.2.1.3 Producer Price of the Crop (Pc)

Producer price of the crop can be approximated either to all India procurement prices or the local selling price of the crops. Since the study tries to find out the cost of soil erosion, which is localized in nature, the local price of the crops, have been considered. Again the average price for three years, *i.e.* 1992-93, 1993-94 and 1994-95 have been taken into account for wheat, rice and maize. This has been given below :

Local Price (Rs./quintal)

Crops	*(Rs. /t)*
Wheat	417.33
Rice *	610.80
Maize	452.00

*The price for rice is the average of fine quality and normal quality.

Source: Sankhikiya Patrika, Dehradun, 1996.

9.3.2.1.4 Rate of Discount (r)

In order to arrive at the present value of cumulative loss in yield of crops, a real rate of discount should be applied. The real rate of discount in the economy implies the real rate of economy-wide return. It is based on the marginal cost of capital approach to the rate of discount. For any specific sector, the rate or discount means the rate of return, *i.e.* the rate of growth of that particular sector. Here the rate of discount to be applied is for agricultural sector in UP state. Hence, rate of return can be approximated to the rate of growth of agricultural sector in UP. This rate of agricultural sector has been around 3.1 per cent per annum from 1980-81 to 1996-97 (Sankhikiya Diary, 1997). Hence a 3 per cent rate of discount has been considered as appropriate. However, for simulation purpose 4 per cent rate of discount has also been considered in the study.

9.3.2.1.5 Time Horizon (n)

Time horizon of 5 years, 10 years and 20 years have been considered. Most of the conservation measures last for 10 years hence these figure commands a special place for cost benefit analysis for soil and water conservation projects.

Now for estimating the production value of soil erosion in the valley, it is presumed that in the entire area of the valley (2,09,000 ha) only wheat or only Paddy or only maize is cultivated in a year. It is unusual but for cost estimation purpose this assumption is not unrealistic and helps us to arrive at reliable estimate.

After substituting the value of relevant variable in the expression

$$J_T = \sum_{n=0}^{N} \frac{P_c(Y_0 - Y_0(1-d)^n)}{(1+r)^n}$$

We get the value or yield loss/ha for year 1994 for wheat, rice and maize separately. Following tables provides the estimate.

Under the assumption that the total area of Doon valley is cultivated for wheat, rice and maize respectively, the one time annual damage comes out maximum when only rice cultivation is presumed which is around Rs. 3427.6/ha for 1994 at current prices at 3 per cent rate of discount. If the rate of discount is taken as 4 per cent, the figure comes out as Rs. 3394.64/ha. The figure for maize cultivation (presuming that only maize is cultivated in entire area) comes out as Rs. 1715.62/ha and Rs. 1699.13/ha at 3 per cent, and 4 per cent rate of discount respectively (Table 9.3). Assumption of wheat cultivation gives the intermediate figure of Rs. 2550.98/ha and Rs. 25216.45/ha for the corresponding rate of discount respectively. Higher figure of loss under the assumption of rice cultivation is due to the higher price much above that of maize and wheat.

Many studies of soil loss has calculated the current annual loss in agricultural output due to topsoil loss. For India studies had done by Brandon *et al.* (1995) and TERI (1997) estimates only one time loss in productivity, which is not very correct. Erosion-induced productivity losses are not limited to a particular period but stretches very far in the future and it accumulates through many years to come. Crosson and Stout (1983) put it as "Erosion-induced productivity losses are not confined to a terminal year, but accumulate over much of all of the intervening period. Consequently knowledge of the effect of degradation on costs in a terminal year is incomplete. The way to compensate for these deficiencies is to calculate the present value of the annual productivity loss."

Table 9.3: Total Production Value of Eroded Soil in Doon Valley in 1994 at Current Price (Rs. crore)

Crops	*Current annual r*		*N = 5 yrs r*		*N = 10 yrs r*		*N = 20 yrs r*	
	3 per cent	*4 per cent*	*3 per cent*	*4 per cent*	*3 per cent*	*4 per cent*	*3 per cent*	*4 per cent*
Wheat	53.3 (2550.98)	52.8 (2526.45)	450.00 (21532.13)	436.00 (21532.13)	980.00 (46908.48)	927.00 (44357.8)	1850.00 (88561.2)	1678 (80314.5)
Rice	71.6 (3427.6)	70.9 (3394.638)	648.5 (31032.6)	627.9 (30047.6)	1470 (70268.2)	1387 (66376.7)	2840 (135962.1)	2572 (123083.8)
Maize	35.85 (1715.625)	35.5 (1699.13)	320 (15350.0)	310 (14864.3)	721 (34517.97)	681.6 (32612.4)	1390 (66478.68)	1258 (60201.34)

Here in estimating the costs of erosion in Doon valley through productivity approach two real rate of discount, *i.e.* 3 per cent and 4 per cent (Planning Commission, Government of India recommends 12 per cent as the nominal rate of discount for Indian context). Albeit an ideal path to derive rate of discount could have been an environmentally adjusted one for the valley as has been recommended in the standard literature of environmental economics, (Markandya and Pearce, 1988). However, these rates of discount will not be far away from perfection. Three possible time horizons of 5 years, 10 years and 20 years have been considered for estimation of present value of accumulated loss.

9.3.3 The Resource Value of Soil Erosion

In order to estimate the loss due to erosion in the valley relies on the accumulated cost of replacement of nutrient lost on account of topsoil lost. In the expression

$$B_T = \sum_{n=0}^{N} \frac{P_1D_1 + P_2D_2 + P_3D_3 + P_4D_4}{(1+r)^n}$$

P_1, P_2, P_3 and P_4 are the prices of nitrogen, potassium, phosphorous and organic carbon obtained through deploying artificial fertilizers.

D_1, D_2, D_3 and D_4 are the nitrogen, phosphorous, potassium and organic matter lost in the valley by n years.

r and n are the real rate of discount and total number of years taken into account.

Prices of Fertilizers (P's): The major nutrients lost have been identified as nitrogen (N), phosphorous (P), potassium (K) and organic matter due to the erosion of soil in the valley. Loss of nitrogen can be replaced by Urea (46 per cent N) although other fertilizer can also replenish the lost nitrogen but here only urea has been considered which contains the maximum concentrations of nitrogen against ammonium sulphate (20.6 per cent N) and calcium ammonium nitrate (25 per cent N). Phosphatous is supposed to be replaced by Diammonium phosphate (18-46-0). Potash has been replenished by muriate of potash (60 per cent K_2O). Prices of fertilizers are generally administered and controlled which varies from state to state depending upon sales taxes and other taxes. Prices of Diammonium phosphate (DAP) and muriate of potash has been decontrolled since August 1992. However, price of urea is still controlled and regulated. In states, seasonal variation in the prices of fertilizers is also prevalent. Disaggregated level data for Doon valley is not available. In UP, prices, for different fertilizers in order to obtain the nutrients come as follows:

Table 9.4: Nutrient Prices in UP in 1994 at Current Price (Rs. /kg.)

Nutrients	*Prices (1994)*		*Average Price*
	Kharif	*Rabi*	
Urea (for N)	7.22	7.22	7.22
DAP (for P)	12.18 to 14.07	13.58 to 16.30	14.03
MOP (for K)	5.94 to 6.50	6.13 to 6.57	6.28
Organic matter*	4.80	5.25	5.00

*Price for organic material has been taken from the price of Farm Yard Manure (FYM) sold in Dehradun in 1994.

Source: Fertilizer Association of India, Fertilizer Statistics, 1996-97.

Nutrient Loss Estimation

Doon valley comprises mainly drainage catchment of Yamuna and Ganga rivers. Both these catchments are subjected to felling, lopping, overgrazing, cultivation and mining activities. The interaction of these activities alters the hydrologic regime and soil-plant-water relationship. This result accelerates peak rates of discharge and soil erosion. Therefore, valuable topsoil cover is lost annually. This further results in nutrient loss from these catchments. Therefore, an estimation of nutrient loss has been attempted in the present study. For the calculation of annual nutrient loss through sediment losses, average concentration of different soil nutrients *viz.* total nitrogen, total phosphorous, total potassium, total calcium and organic carbon in run-off soil (mg g^{-1}) under different landuse categories as per (Pandey *et al.*, 1984) in Doon valley was multiplied with annual soil loss. The concentration of various nutrients in soil is given in Table 9.5 and Table 9.6.

Table 9.5: Total Annual Nutrient Loss (Tonnes/Year) in Run-off Soil in Doon Valley

	1972	*1974*	*1977*	*1979*	*1982*	*1993*	*1994*
Nitrogen	48947.49	48985.01	55689.71	54936.25	53958.21	64659.18	64820.28
Phosphorous	923.11	923.55	1050.86	1034.55	1017.66	1219.20	1190.38
Potassium	179343.44	179513.32	204190.33	201445.54	197834.43	237168.20	241752.11
Calcium	388588.05	389400.76	441717.40	434931.29	426821.35	511119.14	431425.71
Organic Carbon	415557.83	422569.52	462395.07	442558.43	429655.33	407552.21	509328.70

Table 9.6: The Concentration of Nutrients (mg/g) in Run-off in Doon Valley

Landuse	*Nitrogen*	*Phosphorous*	*Calcium*	*Organic*	*Carbon*
Dense Forest	2.32	0.044	8.25	18.75	22.50
Open forest	2.21	0.041	8.26	18.00	22.35
Scrub	2.12	0.040	7.75	17.00	21.25
Cultivation hills	2.65	0.051	9.75	20.50	13.00
Cultivation plains	2.64	0.049	9.70	20.40	13.80
Cultivation with habitation	1.77	0.034	7.70	17.50	16.25

On the basis of the soil loss figure obtained for 1994 (latest year) and the major nutrient loss from the previous table, we arrive at the total major nutrient loss in the valley (20900 ha) as follows:

Table 9.7: Major Nutrient Loss in Doon Valley due to Soil Erosion in 1994 (tons)

Nutrients	*Loss (tons)*
Nitrogen (N)	64820.28
Phosphorous	1190.38
Potassium	241752.11
Organic Carbon	509328.71

Organic carbon loss for year 1994 is compensated by using Farm Yard Manure, which contains around 50 per cent of organic carbon. Since the price of FYM for UP in 1994 was Rs. 2.50/kg. The price of organic carbon comes out as Rs. 5.00/kg.

Using the expression for present value of nutrient loss due to soil erosion at 3 per cent and 4 per cent rate of discount for 5 years, 10 years and 20 years of time period the total costs of erosion has been calculated for the valley. Table 9.7 provides the details.

The cost of erosion on annual basis is Rs. 574.07 crore at current price in 1994. However, the total present value of losses over 5 years, 10 years and 20 years are Rs. 3129.7 crore, Rs. 5471.00 crore and Rs. 9114.8 crore (at 3 per cent rate of discount), Rs. 3129.7 crore, Rs. 5230.3 crore and Rs. 8375.87 crore (at 4 per cent rate of discount) respectively.

Table 9.8: Resource Value of Soil Erosion in the Valley Based on Replacement Cost Approach (Rs.) in 1994 at Current Price

		Cost of Erosion	
Variable	*Current annual*	*(Rs. Million)* *5740.7*	*(Rs. Per Ha.)* *27,467*
N = 5 yrs;	r = 3%	32032.0	1,53,263
N = 5 yrs;	r = 4%	31297.0	149,746
N = 10 yrs;	r = 3%	54710.0	261,770
N = 10 yrs;	r = 4%	52303.0	250,254
N = 20 yrs;	r = 3%	91148.0	436,110
N = 20 yrs;	r = 4%	83758.7	400,759

9.4 Comparison of the Two Approaches

Replacement cost approach to valuation of damage due to soil erosion is rooted in deep ecology approach to environmental economics. It emphasises the cost of all lost nutrients replenished through artificial fertilizers should be utilised by the crops or taken into account by the farmers. Therefore, replacement cost approach or resource value of soil is a comprehensive estimation of loss of erosion and is generally taken as the higher range of cost of erosion. On the other hand production value of soil approach tries to capture the marginal decline in the yield due to erosion and is closer to the realised experience. Since plants cannot use all the nutrients available in the soil. In fact large number of experimental data show that only a fraction of nutrients available in the soil ecosystem is used by the plants, productive value of soil erosion is always less than the resource value of soil erosion. Table 9.9 shows the comparison.

While the current loss due to erosion under resource value approach is 10.38 times of the production value approach, the divergence in the cost estimation narrows if longer time horizon is incorporated at all rate of discounts. Over 5 years, the cost of erosion under resource value approach is 7.11 times that of the production value approach at 3 per cent rate of discount, but at 4 per cent rate of discount it becomes 7.17 times. In 10 years the resource value is 5.86 times of production value at 3 per cent rate of discount but at the same rate of discount it comes to 4.93 times for 20 years of period in consideration. Similarly, at 4 per cent rod the resource values for 10 years becomes 5.64 times of the production value and in 20 years 4.99 times of production value. These findings are similar to studies conducted by Stocking for Zimbabwe in 1986 and Clark *et al* for Parawella area of Sri Lanka in 1998. In both the studies only three major nutrients, *i.e.* N, P and K have been taken into account. Stocking considers value of only current annual loss while Clark *et al* takes the present value for 20 years. If organic carbon is not considered then the resource and production value will converge in the long run.

Table 9.9: Comparison of Costs of Soil Erosion Under Production Value and Resource Value Approaches

Variable	*Costs (Rs.)*		*RV/PV*
	Production value (PV)	*Resource value (RV)*	
Current annual loss	5.53×10^{8}	57.407×10^{8}	10.38
3 per cent	4.5×10^{9}	320.97×10^{8}	7.11
n = 5 yrs x=			
4 per cent	4.36×10^{9}	312.97×10^{8}	7.17
3 per cent	9.8×10^{9}	574.03×10^{8}	5.86
n = 20 yrs x=			
4 per cent	9.27×10^{9}	523.03×10^{8}	5.64
3 per cent	1.85×10^{10}	911.48×10^{8}	7.93
n = 10 yrs x=			
4 per cent	1.678×10^{10}	837.587×10^{8}	4.99

As mentioned earlier, plants take only a fraction of nutrients available in the soil, resource value of soil exceeds the production value. But only nutrients are not the essential elements provided by soil to the plants, water-holding capacity of soil plays a significant role in the growth of crops. Other factors like living organism etc. are also lost when the soil is eroded. Hence a comprehensive resource value of soil will be much higher than the replacement of major nutrients by artificial fertilizers. Further the erosion-productivity relationship has been extended for the whole Doon Valley on the basis of data obtained from laboratory. Ideally for every class of slope, the impact of soil loss on crop's productivity will vary and hence rigorous and extensive scientific data is needed to arrive at some more meaningful conclusions about the costs of erosion on these two approaches.

10

Soil Conservation in Decision-Making Framework

In this chapter, appropriate conservation measures for reducing soil loss shall be analysed. This will be done on the basis of physiographic analysis of Doon valley region with the help of GIS techniques. Then different physical requirements of labour and capital will be guaged, the cost of conservation measures relevant in the valley with optimal efficiency will be estimated. Finally, cost-benefit analysis of soil conservation in the given valley area shall be done to examine the rationale of investible funds meant for conservation programme.

10.1 Soil Conservation: Areas and Measurement Costs

Soil conservation is a remedial measure, which are generally adopted in the realised experience of excessive erosion. It is true for areas occupied by poor farmers who go for it when the problem seems to be acute in nature. Every bit of soil erosion can neither be checked entirely nor it is derivable. For Doon valley too, the entire area has been categorised into various erosion class depending upon the level of soil erosion. The existing and potential erosion prone areas were delineated from satellite imagery (FCC) of 1994 on the basis of tonal, textural and physiographic recognition of terrain features like change in soil colour, appearance of bare soil/rock/gravel etc., accentuated dendritic drainage pattern and land-use change.

Following table provides the erosion classes—area information for Doon valley.

Table 10.1: Area-Erosion Classes for Doon Valley in 1994

Erosion Class (t/ha/y)	*Area*
Nil to slight (up to 5)	71,595
Slight to moderate (6 to 10)	52,800
Moderate (11-25)	9,378
Severe (26-50)	48,417
Very Severe (51 and onwards)	26,810
Total	**20,9000**

For the present cost estimation the areas under severe and very severe erosion classes have been considered for conservation purposes. This 75,227 ha of area, which is around 36 per cent of total Doon valley area, causes the maximum soil loss. Remaining areas albeit need attention have not been considered here because priority setting criteria implicitly favours the conservation of most threatened area first, *i.e.* conservation of soil falling under severe and very severe erosion class.

There are various measures to check the erosion of topsoil depending upon the physiography of the land and financial incentive available for the owners. The two main categories of soil conservation measures are mechanical and biological one. Both the measures differ in their application, effectiveness, cost structure and duration. Mechanical measures also known as engineering measures usually involve creating barriers across the direction of the flow of water and thus retard the runoff and subsequently soil loss. Such measures generally include contour cultivation, contour bunding, graded bunding, gully control, stream bank protection, bench terracing, trenching etc. Biological measures include grass plantation, horticulture and development of vegetative cover on the denuded fragile land. These biological measures take longer time to be effective but they inflict lesser damaging effect on the soil than the mechanical one. Both the measures can be used in ideal combination on a piece of land and hence they are complementary to each other. These measures are not two alternatives but are complementary to each other. Mechanical measures provide foundation and skeleton to the eroded area while biological measures act as living and production organs of that region. Mechanical measures provide immediate check to the erosion but generally poor farmers are reluctant to adopt mechanical measure due to higher initial investment.

The development of a soil conservation strategy has to be based on a number of physical characteristics. Different conservation strategy and their choice for checking the soil loss depend upon the suitability and capacity of the land. The suitability of geography units can be defined as a measure of the physical capacity of specific location to support a specific landuse. This is based upon physical factors, describing the characteristics of the watershed and its landuse—slope, soil type, vegetation, meteorological and hydrological factors, etc. The main objective of conservation measures is to check or reduce the run-off soil loss in Doon valley catchment area. For this purpose a number of studies conducted earlier have been taken into consideration, especially to evaluate the cost effectiveness of these conservation measures and which, have proved as efficient. These studies, however, were restricted to very small areas, *i.e.* micro-watersheds in the valley.

We have attempted to extend the application of these conservation measures to the entire Doon valley. The data from these studies has been liberally used to arrive at a logical framework for evolving a suitable mix of different strategy in the valley. The first step in this direction was the preparation of soil erosion map of the valley (Fig. 8.2), which was prepared with the help of scientists of CISMHE on the basis of slope classification, physiography, soil, land-use and drainage pattern. This soil erosion map was prepared by using Geographical Information System (GIS) overlay analysis using Arc View GIS 3.0a. It allows for the assignment of specific land-use activities to the most suitable available land units, based on appropriate measures of suitability. GIS are computer based (hardware and software) tool to record, analyse, represent and describe area spatial information in two-dimensional or three-dimensional manners.

After preparing the map of soil erosion for the valley, areas needing different conservation measures were identified in close association with the hydrologists and soil scientists of CSWCRTI and Wadia Institute of Himalayan Geology both located in Dehradun. Following Table 10.2 provides the information.

The areas with very steep slopes were excluded due to their inaccessibility. Only the areas where cultivation is not suitable have been considered for biological measures. The extent of area earmarked for different treatment measures is likely to vary as the estimates were calculated from maps on 1:50,000 scale wherein entire

Doon valley with an area of about 2090 sq km is required to be sub-divided into a number of sub-watersheds and micro-watersheds ranging from an area of thousand hectares to a few hectares. This requires micro-level delineation of watersheds and maps up to 1:1000 scale needs to be prepared. Therefore, there is bound to be a variation in extent of areas earmarked for various conservation measures. Similarly the cost estimates would also have similar errors. However, here an attempt was made to arrive at some figure keeping all the limitations in mind. The areas with very severe erosion with steep slopes and which are inaccessible were not considered for any conservation measures and are being recommended to be left alone for natural rejuvenation. Various mechanical (engineering) and biological conservation measures to be adopted in Doon valley have been described below.

Table 10.2: Areas for Different Conservation Measures

	Conservation Type	Area (in ha)
Mechanical Measures		
1.	Gully Control	11,031
	— DRSM check dams (Model I)	4,930
	— Combination of DRSM and crate works (Model II)	5,520
	— Combination of DRSM check dams, check walls and crate works (Model III)	581
2.	Bench Terracing / Contour trenching	7,200
3.	Stream Bank Protection	12,650
4.	Landslide Control Measures	400
	Total	**31,212**
Biological Measures		
1.	Degraded areas restoration	40,750
2.	Pasture development	1,890
	Total	**42,640**
	Grand Total	**73,921**

10.1.1 Mechanical Measures

10.1.1.1 Gully Control

Gullies are formed mainly on account of physiography, soil type and heavy biotic interference in the area. The scouring of streams like various nallahs in Siwalik and Doon valley regions at their peak flows with sediment-laden run-off and it causes gully formation. These areas required to be closed to all biotic interference and construction of contour bunds/check-dams/peripheral bunds and also the gully plugging are strongly advised. All these measures promote the growth of vegetation leading to stabilization of slopes/areas.

In Doon valley, different types of gully plugging/check dams are sought to be employed for different conditions depending upon the availability of type of material and gully width. Gully plugging by construction of brushwood dam has not been recommended due to high cost in the long run and susceptibility to white ant attack. Earth is the cheapest and most readily available material and therefore, it is most economical to construct the earthen gully plugs. In Doon valley composite check dams comprising earth in the non-flow sections and brick/stone masonry in the spillway portion have been suggested. Depending upon the width of the gullies three different models have been recommended.

The following types of models under gully control measures are recommended.

Model I DRSM (Dry Rubble Stone Masonary) Check dams with the stone available at site.

Model II Combination of DRSM and crate works. These are recommended for moderate to deep gullies with stones available at sites.

Model III Combination of DRSM check dams, check walls and crate works. These are recommended for areas with eroding hill slopes.

Under gully control measures, different combinations of DRSM check dams, check walls and wire crates are used for shallow, moderate, main and side gullies. These combinations have been named as Model I, Model II and Model III. For Model I, the average dimension of check dams is 8m × 1.5m × 2m = 24 cubic metre (cum). The cost of construction of check dam for this size given by the Forest Department of UP State is Rs. 180 per cum at 1998 price; hence the total cost for single check dam comes out as Rs. 4,320. The area of 5 ha facing gully erosion will need at least 8 check dams. Therefore, the cost of check dams for 5 ha area will be Rs. 8 × 4,320 = Rs. 34,560. Accordingly the cost per ha comes out

Rs. 6,912 at 1998 prices. The detailed calculation has been given in Table 10.3.

Similarly under Model II, the cost for labour and dam construction materials comes out as Rs. 9,300/ha. Cost estimates for different minor materials have been used from state forest department of UP.

For Model III, the cost comes out as Rs. 11,590/ha. The details have been worked out in Table 10.3.

Table 10.3: Cost Estimate for Gully Control Measures

MODEL I: DRSM check dams

Conditions

- Gullies: shallow to moderate
- Stones available at site locally

Details for a unit area of 5 ha

No. of check dams to	=	8 Nos. be erected
Average dimension	=	(8m × 1.5m × 2m)
	=	24 cum
Cost @ Rs. 180/cum	=	24 × 180
	=	Rs. 4,320 per no.
Average cost for 8 Nos.	=	Rs. 34,560
Cost per ha	=	34,560/5 = **Rs. 6,912**

MODEL II: Unit cost of check daming/gully plugging works (Per ha)

Conditions:

- Main gullies — Crate & DRSM
- Side gullies — DRSM check dams
- Stones available at site (To be extracted and collected locally)

Details worked for a unit area of 20 ha

(i)	No. of crates to be erected	=	9 Nos.
	Average dimension	=	(8m × 1.5m × 2m)
		=	24 cum
	Total stones required	=	216 cum
	(a) Cost of crate wire @ Rs. 2650 per qtl.	=	Rs. 23,850
	(b) Carriage, weaving, etc., L.S.	=	Rs. 3,500
	(c) Cost of crate with local stones available including tie works	=	216 × 180 = Rs. 38,880
	Total cost (a + b + c)	=	Rs. 66,230

(ii)	Total No. of check dams inside gullies	=	40 Nos.
	Average dimension	=	(5m × 1.5m × 1.5m)
		=	11.25 cum
	Total stone required	=	450 cum
	Construction cost with local stones available @ Rs. 180/cum	=	Rs. 81,000
(iii)	No. of check dams in main nala	=	9 Nos.
	Average dimension	=	(8m × 1.5m × 2m)
		=	24 cum
	Total stone required	=	216 cum
	Construction cost with local stones including tie works @ Rs. 180/cum	=	Rs. 38,880
	Therefore *total cost/ha* (i + ii + iii)	=	1,86,110/20
		=	Rs. 9,305/ha
		Say	**Rs. 9,300/ha**

MODEL III: Unit cost of check dams, check walls and crate works (per ha)

Conditions:

– Combination of DRSM check walls, check dams and crate work in eroding hill slopes

Details worked for a unit area of 20 ha

(i)	No. of check walls to be erected	=	50 Nos.
	Average dimension	=	(10m × 0.8m × 0.5m)
		=	4 cum
	Total stones required	=	4 × 50 = 200 cum
	Cost of checkwalls with locally available stones to be extracted and collected @ Rs. 180 per cum	=	Rs. 36,000
(ii)	No. of check dams to be erected	=	30 Nos.
	Average dimension	=	(8m × 1.5m × 2m)
		=	24 cum
	Total stones required	=	24 × 30 = 720 cum

	Construction cost @ Rs. 180/cum with locally available stones to be extracted and collected	=	720 × 180
		=	Rs. 1,29,600
(iii)	No. of crates to be erected	=	9 Nos.
	Average dimension	=	(8m × 1.5m × 2m)
		=	24 cum
	Total stones required	=	24 × 9 = 216 cum
	Cost of crate wire for 9 qtls. @ Rs. 2650/- per qtl.	=	Rs. 23,850
	Carriage weaving, etc.	=	Rs. 3,500
	Cost of crate work with locally available stones including collection and extraction @ Rs. 180/cum	=	Rs. 38,880
	Total cost	=	Rs. 66,230
	Therefore, *total cost/ha* (i + ii + iii)	=	Rs. 2,31,830/20
		=	Rs. 11,591

Say Rs. 11,590 per ha

Note: All figures are in 1998 prices. The rates for different wage categories have been literally taken from Forest Department, UP state.

10.1.1.2 Stream Bank Protection

Stream bank erosion is widely prevalent in eastern and western part of Doon valley and is caused by destruction of vegetative cover, mass movement on unstable bank slopes, toe cutting, sliding of slope when saturated with water. For this situation, the appropriate control measures include construction of wire crates and vegetative spurs. All the streams in Doon valley affected by this phenomenon are to be taken up for these conservation measures.

A wire crate of 2m × 2m × 15m need labour for site preparation and stone collection of mandays for each activity. Cost per structure comes out to Rs. 9,990. For vegetative spur, main items required are brushwood, wooden poles and labourers. 40 bundles of brushwood in two layers need 16 poles of 6′ × 6′ with 6′ diameter and 4 labourers. The total cost comes out as Rs. 1200 per structure. Table 10.4 provides the details.

Table 10.4: Cost Estimate for Stream Bank Protection Measures

Sl. No.	*Item*	*Unit*	*Qty.*	*Rate (Rs./unit)*	*Amount (Rs.)*
A.	**Wire Crates: Size (2 m × 2 m × 15 m)**				
1.	Crate wire	Cum	60	75.00	4,500.00
2.	Labour site preparation	Mandays/ cum	60	45.75	2,745.00
3.	Stone collection fill & tie	Mandays/ cum	60	45.75	2,745.00
		Total (Rs.)			9,990.00
	Therefore, Cost per structure = Rs. 9,990/-				
B.	**Vegetative Spur**				
1.	Supply of brushwood material with 40 bundles in two layers	Per Bundle	40	15.00	6,00.00
2.	Cost of wooden poles 6′ × 6′ and 6′ dia	Nos.	16	25.00	4,00.00
3.	Labour charges for laying of brushwood in spur and fixing of spur	Labourers	4 Nos.	50.00	2,00.00
	Total (Rs.)				1,200.00
	Therefore, Cost per structure = Rs. 1200				
	All prices are of 1998.				

10.1.1.3 Landslide Control

The landslide prone areas primarily on the existing road length in Doon valley are required to be taken up for following conservation/control measures.

(i) Breaking of slope by wattling, retaining wall, logwood checks, narrow based terraces, etc.
(ii) Control measures to stabilize the main drainage channel.
(iii) Vegetative methods

(a) grasses,

(b) shrubs and trees,
(c) mulching.

The detailed cost estimates of these treatment measures could not be worked out separately. However, based upon similar experiences in Doon valley and elsewhere, rough estimates of cost of treating landslides per hectare have been adopted for this study. The average cost for landslide control comes out as Rs. 10,000/ha (Department of Forest, UP, 1998).

10.1.1.4 Bench Terracing/Contour Trenching

Bench terracing is recommended in areas with steep slopes in Doon where cultivation is practiced on hill slopes. This involves reduction of slope length and modification of degree of slope in order to reduce the intensity of scouring action of run-off water. In this technique, the area is converted into level step-like fields constructed by half-cutting and half-filling. This helps in reducing the degree of slope substantially. It also helps in the uniform reduction of soil moisture, retention of soil and manure and in efficient utilization of irrigation water where applicable. Contour trenching is also recommended for cultivation areas with 10-16 per cent slope to reduce the run-off and soil loss.

The detailed cost estimates for bench terracing per unit area could not be worked out as the cost of construction of bench terraces depends upon steepness of slope and the length of slope to be treated. However, an average cost estimate per hectare has been taken in the study from the Department of Forest, UP (1998) as Rs. 8,000/ha.

10.1.2 Biological Measures

The total areas requiring biological conservation measures in Doon valley was estimated from the extent of area under forests with open canopy, scrub and barren areas in the valley in different slope categories.

The plantation of trees and grasses is the economically best-suited conservation measures to be adopted for areas, which are not suitable for agriculture. The vegetation canopy thus developed is very effective in intercepting rainwater and also reduces the energy of raindrops falling on soil and the soil dispersion. In Doon valley, rising of plantations of locally useful indigenous plant species are recommended based upon site-specific conditions. In addition to afforestation of degraded areas in Doon valley, conservation horticulture is another measure where cultivation is not profitable. Horticultural trees not only provide the protection to degraded lands but act as a stable source of income. This approach has already been adopted extensively by CSWCRI, Dehradun in Doon valley. The raising of fuelled/fodder plantations provides another sustainable land management system as a part of agro-forestry programme.

For all activities under biological measures, labour remains the dominant requirement. Right from digging of pits and bush cutting to refilling of pits, weeding, raising seedlings, carriage, planting, cutting fencing, stretching and fixing of barbed wire needs semi-skilled and agricultural labour. Cost of some materials like barbed wire etc. is not very high. Another significant component of cost include maintenance cost for plantations, which accrues over time and has been excluded here since only initial cost of conservation is being estimated in this study. Following table gives the detailed information.

Now after deriving the individual cost figures for each conservation types to be adopted in the valley, total cost has been calculated by multiplying with the corresponding area. Table 10.6 gives the detail.

Table 10.5: Restoration of Degraded Areas: Plantations Cost Per Hectare (1500 plants/ha)

Sl. No.	*Item of Work*	*Cost (Rs.)*
A.	**Wage Component**	
A1.	Bush cutting & cleaning of site @ Rs. 364 per ha	364.00
A2.	Digging of pits (45cm × 45cm × 45cm) @ Rs. 125/100 Nos.	1,875.00
A3.	Refilling of pits @ Rs. 85 / 100 Nos. pit	1,275.00
A4.	Weeding & hoeing once @ Rs. 25 / 100 Nos.	375.00
A5.	Cost of raising seedlings in nursery @ Rs. 200/100	3,000.00
A6.	Carriage of plants from nursery to site an Average of 4 km @ Rs. 38.00 / 100 plants/km	2,280.00

A7.	Planting of seedings @ Rs. 61/100 Nos.	915.00
A8.	Cutting & preparation of fence posts 80 Nos. @ Rs. 395/100	316.00
A9.	Coal taring of fence posts @ Rs. 85.00/100 Nos.	68.00
A10.	Carriage of fence post over a distance of 2 km @ Rs. 2100/100 Nos. per km	336.00
A11.	Digging of 45 cm deep holes for fixing of fence posts 80 Nos. @ Rs. 150/100 Nos.	120.00
A12.	Fixing of fence posts 80 Nos. @ Rs. 225/100 Nos.	180.00
A13.	Stretching and fixing of barbed wire in 4 strands total 960 rmt @ Rs. 1.50/rmt	1,440.00
	Total (A)	**12,544.00**
B.	**Cost of material**	
1.	Cost of barbed wire approx. 80 kg. @ Rs. 25/kg	2,000.00
2.	Carriage of Barbed wire upto the site for an average distance of 2 km	200.00
3.	Cost of U-nails L.S.	50.00
	Total (B)	**2,250.00**
	Therefore, Total cost of plantation per ha (1500 plants) (A + B)	**Rs. 14,794.00**
		Say Rs. 14,800 per ha

Note: The total cost for biological measures comes as Rs. 14,800/ha.

Table 10.6: Cost Estimates for Conservation Measures

Sl. No.	*Item of work*	*Unit*	*Qty.*	*Rate (Rs. /unit)*	*Amount (Rs.)*
1.	Gully control measures	Ha	11,031		
	Model I	Ha	4,930	6,912	3,40,76,160
	Model II	Ha	5,520	9,300	5,13,36,000
	Model III	Ha	581	11,590	67,33,790
2.	Steam bank protection	Ha	3,650		
	(a) Wire crates	No.	3,540	9,990	3,53,64,600
	(b) Vegetative spurs	No.	10,490	1,200	1,25,88,000
3.	Landslide control measures	Ha	400	10,000	40,00,000
	(a) Construction of retaining wall				
	(b) Stepping/terracing of slope				
	(c) Vegetative spurs				
4.	Bench terracing/Contour trenching	Ha	12,650	8,000	10,12,00,000

5.	Restoration of degraded areas (a) Afforestation (b) Fueld/Fodder Plantation (c) Horticultural plantations	Ha	40,750	14,800	6,03,10,000
6.	Pasture development	Ha	1,890	5,000	94,50,000
7.	Total				85,78,48,550

Rs. 85.785 crore.

Thereby the cost of conservation with 75-80 per cent efficiency rate for checking the erosion of 73921 ha of area in Doon valley comes out as Rs. 85.785 crore at 1998 price.

10.2 Benefit Cost Analysis of Conservation Measures

In spite of several limitations and criticisms, Benefit-Cost Analysis still remains the dominant tool of any decision making framework. Going by the efficiency criteria, CBA provides a very useful thumb rule to decision makers. Ideally, benefit-cost analysis would involve the appraisal of all costs and all benefits to a proposed course of action or investment project from the viewpoint of society as a whole. Benefits and costs depend upon the vector of objectives laid down in the formulation. Assigned objectives function as a reference point and around this reference, benefit avoided becomes cost and cost avoided becomes benefit. For Doon valley, soil loss prevented with the help of conservation measures is the benefit and cost incurred in conservation efforts is the cost. Further, benefits or costs have been considered from the local farmer's point of view. This is primarily due to the reason that decisions about land use and soil conservation practices are ultimately taken by the farmers themselves in light of their own objectives, production possibilities and constraints and not by social planners or government agencies. Moreover the benefit-cost analysis should be done at local and micro level because soil conservation problems are site-specific. Cost-benefit analysis techniques offer a simple but powerful tool to the analysis of soil conservation problems. Cost-benefit analysis technique was originally developed as a tool to examine the economics of projects involving the development of water resources, but in last three decades it has gained prominence for efficient management of natural resource (Krutilla and Fisher, 1975; Hufschmidt *et al.*, 1983). Various variants of this technique has been used to examine the economic efficacy of soil conservation strategies for example for India (Magrath, 1989), the Dominican Republic (Veloz *et al.*, 1985), Mali (Bishop and Allen, 1989) and several cases in Kenya (Hedfors, 1981; Holmberg 1985; Lindgran, 1988; Pagiola 1992).

In the present scheme, the benefit of soil conservation has been approximated to the value of soil loss following the resource value approach. Production value of soil for severe and very severe erosion class is extremely difficult to estimate. Relationship or yield decline and degree of erosion becomes absolutely vague in absence of elaborate Experimental data for agricultural field of every slope category. Hence value of lost nutrient mainly N, P, K and organic matter has been considered for calculation of benefits of conservation measures.

10.2.1 Benefit of Soil Erosion (or Prevented Soil Loss)

As mentioned in the previous chapter, areas falling under severe and very severe erosion class are 48417 ha and 26810 ha respectively, totalling to 75,227 ha in the valley. Out of this 75,227 ha, 1306 ha area has been excluded because this area is either inaccessible or are at very steep slope (> 45°). The areas of erosion class have been interpreted from the erosion map for the valley pertaining to year 1994. The conservation measures have been designed for 73,921 ha of area hence the benefits of the conservation measure will be the prevented soil loss due to these measures. Appendix I gives the details of soil loss for each land use category and the erosion map provides the area for each land use category. Following table provides the detailed information:

In the entire Doon valley for 2,09,000 ha of area the soil loss was around 27.9 mill. tons but the area under severe and very severe erosion class, *i.e.* 75,227 ha experiences the soil loss of 15.2 mill. ton. In other words, 36 per cent of total area of the valley looses around 55 per cent of the total loss. Therefore the conservation measures have been sought for such areas only. However, the excluded region of 1,306 ha primarily fall under severe erosion category and contributed around 46,076.28 tons (1,306 × 35.28) of soil loss. Here the effective soil loss remains at 15.2 mill. tons in 1994.

Table 10.7: Soil Loss (T) For Severe and Very Severe Erosion Category in 1994 in Doon Valley

Erosion Category	*Land-use/cover*	*Area (ha)*	*Erosion rate (t/ha)*	*Total soil loss (L)*
	Sparse forest	263	34.61	9,102
	Open forest	7,483	35.28	2,64,000
	Scrub	18,954	47.78	9,05,622
	Medium to high cultivation	13,014	35.92	4,67,463
Severe	High intensity cultivation	8,703	40.70	3,54,212
	Sub-Total	48,417		20,00,399
	Old terrace cultivation/ barren	2,315	764.4	17,69,586
	Barren/scrub	5,060	6,21.0	31,42,260
Very Severe	Terrace cultivation	14,066	5,76.0	81,02,016
	Open forest	5,369	54.6	2,93,147
	Sub-Total	26,810		13,207,009
	Grant Total			15.2 mln. tonnes

The value of lost major nutrients, *i.e.* N, P, K and organic matter have been estimated by using the resource value expression, *i.e.*

$$B_T = \sum_{n=0}^{15} \frac{P_1D_1 + P_2D_2 + P_3D_3 + P_4D_4}{s(1-r)^n}$$

where

P_1 = price of nitrogen (Rs. /kg.)

P_2 = price of phosphorous (Rs. /kg.)

P_3 = price of potassium (Rs. /kg.)

P_4 = price of organic matter (Rs. /kg.)

D_1 = loss of N (in kg)

D_2 = loss of P (in kg)

D_3 = loss of K (in kg)

D_4 = loss of organic matter (in kg)

All figures pertain to year 1994. n = Time horizon for the conservation measures. Almost every mechanical measures remain effective for at least 15 years. Mechanical measures adopted for soil conservation prove to be effective in controlling on average 75-80 per cent soil loss in the valley (Dhruvnarayana, *et al.,* 1992). But biological measure provides 70-80 per cent effectiveness for 7-8 years, therefore, for biological measure investment requirement doubles in 15 years especially for restoration of degraded areas and pasture development (Directorate of UP Watershed Management, Dehradun, 1993). Hence, the time horizon is 15 years.

The concentration of different major nutrients has been given in Chapter 8. And since the total loss of soil from severely and very severely eroded area is around 15.3 mill. tons. Hence, the amount of lost nutrient will be as follows:

Nutrients Loss (in crore kg)

N	2.650
P	0.048
K	9.878
Organic Matter	20.811

The price of nutrients (Rs./kg) has been given in Chapter 8, *i.e.*

$P_1 = 7.22$
$P_2 = 14.03$
$P_3 = 6.28$
$P_4 = 5.00$

The real rate of discount is 3 per cent.

After substituting these values we get the value of soil loss during 15 years, which in turn is a benefit if it is saved. Following are the value of the benefits.

10.2.2 Cost of Conservation (or Cost of Conservation Measures)

Cost of conservation measures have been estimated as Rs. 85.78 crore at 1998 price. In order to calculate the cost at 1994 price, this estimate has been deflated by the corresponding wage index of the agricultural labour in UP state (please see Appendix II). All conservation measures biological and mechanical primarily require labour (around 90 per cent) and remaining items include stone, grass etc. Hence, the deflation coefficient of 0.81 has been derived from the wage index of agricultural labour. Now the cost of conservation at 1994 price comes as Rs. 69.48 crore (0.81 × 85.78 crore). This is the initial cost of conservation measure in the current year. Mechanical measures applied in the valley for the given area will need proper maintenance and administration costing 1.491 crore every year (Deptt. of Forest, UP, 1998). Biological measures will also need maintenance costs in the following way:

Year	*Benefits (in Rs. crore at 1994 price)*
1	180.48
2	175.22
3	170.12
4	165.16
5	160.35
6	155.68
7	151.15
8	146.74
9	142.47
10	138.32
11	134.29
12	130.38
13	126.58
14	122.89
15	119.31

Cost figures have been deflated at 1994 price (wage index) and the total cost has been calculated for the area of 40,750 ha where biological methods are supposed to be adopted. The biological method is repeated after 7 years.

Years	*Cost (1998)*	*Total Cost (Rs. crore at 1994 price)*
1st yr.	Rs. 2000/ha	6.60
2nd yr.	Rs. 1600/ha	5.28
3rd yr.	Rs. 1200/ha	3.96
4th yr.	Rs. 1000/ha	3.30
5th yr.	Rs. 800/ha	2.64
6th yr.	Rs. 800/ha	2.64
7th yr.	Rs. 800/ha	2.64

10.2.3 Benefit-Cost Analysis

Following table provides the information on benefits and costs for the 15 years of time period.

Table 10.8: Benefit and Cost of Soil Conservation Programme in the Valley

Year	*Benefits (in Rs. crore)*	*Cost (Rs. crore)*		*Total cost (Rs. crore at 1994 price)*
		Mechanical	*Biological*	
0	0	19.80	49.61	69.48
1	180.48	1.49	6.60	8.09
2	175.22	1.49	5.28	6.77
3	170.12	1.49	3.96	5.45
4	165.16	1.49	3.30	4.79
5	160.35	1.49	2.64	4.13
6	155.68	1.49	2.64	4.13
7	151.15	1.49	2.64	4.13
8	146.74	1.49	49.61	51.10
9	142.47	1.49	6.60	8.09
10	138.32	1.49	5.28	6.77
11	134.29	1.49	3.96	5.45
12	130.38	1.49	3.30	4.79
13	126.58	1.49	2.64	4.13
14	122.89	1.49	2.64	4.13
15	119.31	1.49	2.64	4.13

*In the 8th year, the investment in biological measure is repeated.

For appraising the economic feasibility of the conservation measures, following criteria is used:

10.2.3.1 Net Present Value (NPV)

This is the method where the costs and benefits generated by a project over time are discounted and then summed to obtain an overall figure. If this figure is greater than zero, then the project in question becomes feasible, otherwise it fails, indicating that it does not generate discounted benefits large enough to offset discounted costs.

$$NPV = \frac{(B_0 - C_0)}{(1+r)^0} + \frac{(B_1 - C_1)}{(1+r)^1} + \ldots.. \frac{(B_n - C_n)}{(1+r)^n}$$

where

t = time, 0, 1, 2, ..., n

n = project's life

B_t = Benefit at time t

C_t = Cost at time t

R = discount rate

For the social conservation project, benefits and costs are given in the table 10.8

$n = 15$ years
$r = 3$ per cent

Therefore,

$$NPV = \text{Rs. } 1624.44 \text{ Cr.} > 0$$

Since the value of NPV far exceeds zero, the conservation measures seem to be quite lucrative and the whole project is worth taking.

10.2.3.2 Benefit-Cost Ratio

It is the ratio of the project's discounted benefits to discounted costs which must be greater than one in order to make the project as feasible.

$$BCR \quad \sum_{t=0}^{n} \frac{1}{(1+r)^t} B_t$$

$$\sum_{t=0}^{n} \frac{1}{(1+r)^t} C_t$$

Here for soil conservation venture in Doon valley

$$\sum_{t=0}^{n} \frac{1}{(1+r)^t} B_t = 1791.24$$

and

$$\sum_{t=0}^{n} \frac{1}{(1+r)^t} C_t = 170.45$$

Hence BCR = 10.50 > 1

Therefore the conservation is very much desirable.

This benefit cost ratio for different conservation measures for the valley may appear on the higher side but it is justified because we have considered only severely and very severely eroded areas in the valley. It may also be mentioned that in many other cases the benefit-cost ratio has been estimated at almost the same level if not more. Table 10.9 highlights some of the similar studies for evaluation of soil conservation programmes in different parts of the country where the benefit cost ratio has been found unusually high.

Table 10.9: Benefit-Cost Analysis of Soil Conservation Programmes in India

Watershed	*Project Life*	*Discount Rate*	*BCR*	*NPV (Rs. Crore)*	*Source*
Ramganga (UP)	45	10	11.36	120.15	Babu & Dhyani, 1996
Sutlej (Punjab)	45	10	13.56	4835.3	
Mandvarsa (MP)	20	15	9.97	1264.9	
Tejpura (UP)	10	10	9.42	-	KR Dhandapani, 1996

It has been found that as the size of watershed increases the relevant BCR or NPV also increases possibly implying the economies of scale in operation of soil conservation project.

Here it should be noted that benefits have been approximated with the loss of four nutrients only. Soil provides the moisture and other nutrients, which have been ignored here. Besides, many other side benefits have also been ignored like generation of employment through conservation efforts, saved biotic resources in the soil through conservation etc. Needless to add, off-site impacts have been grossly ignored due to the paucity of Experimental and reliable database. In spite of very narrow versions of benefits considered the conservation measures amply demonstrates its economic efficiency on the basis of both the appraisal criteria.

In spite of the feasibility of soil conservation project in the valley, the ground reality is discouraging. Few conservation measures have been applied in practice and even today soil loss is taking place uninhibitedly. The reasons for this are technological and institutional ones. Required technology for soil conservation is either unavailable or alien to the farmer's attitude and conditions. Institutional bottleneck mainly operates through skewed and absentee land ownership pattern.

Generally, the farmers are not aware of the degree of erosion operating over their farmland. Further, farmers are hardly aware of the soil conservation techniques except a few traditional bunding which cannot prevent the erosion on the steep slope. Moreover, knowledge about erosion control measure is a kind of 'public good' where the use of a piece of information relating to a profitable soil conservation measure by one land user does not diminish the supply of that information available for use by other land users. Here emerges a strong need for government backed conservation efforts and investment. These efforts are either absent or have resulted in the technologies of erosion control, which is not welcomed by the farmers due to their conservative outlook, and traditional style of farming.

Soil conservation should be viewed as an investment, both from the perspective of the individual and land user and of the community as a whole. Farmers will go for farming practices and invest in levels of conservation, which are optimum for both them and the community only when there is no market failure or government failure. Generally, these failures become a rule than an exception and farmers base their land use and soil conservation decision on distorted price signals and the resulting allocation of resources into fewer resources for soil conservation measures and the ceases to be controlled. And this process erodes the very base of our agriculture in the valley.

11

Conclusions

Soil is one of the fundamental natural resources. Growing demand for food through cultivation crucially depends upon this resource. This precious natural resource is dwindling because of its erosion. Soil erosion, natural or man-made, becomes serious when the rate of its erosion exceeds the critical limit of its formation. Soil erosion is the most dominant form of land/soil degradation.

Soil erosion has primarily been discussed and researched by the natural scientists (*e.g.*, geologists and botanists). Measure to check the erosion has been developed by the engineers. Economic dimension to the problem of soil erosion has either been missing or has been dealt perfunctorily by the economists. The economists of 18th and 19th century did not address this issue adequately. During their time, industrial revolution had just begun and demand for food was met by traditional cultivation pattern, which did not erode the topsoil excessively. Traditional agriculture had enough flexibility to incorporate the conventional wisdom of farmers who adopted proper conservation practices and kept proper fallow time and to a large extent avoided monocropping. All these were possible because population explosion was yet to occur and food security was well within control.

Adam Smith was a typical product of this context and he was optimist about the future of mankind. He could translate a few technological revolutions of his time into general conclusions, that land resources will never be a problem. Today's problem of widespread soil loss and subsequent land degradation definitely abandons Smith's view and his optimism. Malthus, Ricardo and Mill forecast the doomsday of mankind on the basis of their diminishing returns rule for land resources. However, Malthus mentions about absolute limit of arable land while Ricardo emphasises relative scarcity, which will get reflected through the rising price. Malthusian time-horizon was hazy and Ricardo's belief of roaring prices as a result of relative scarcity of fertile land proved as wrong. Mill's argument was too philosophical to be translated into simple economic analysis at least during his lifetime. Karl Marx gave an entirely different colour to the problem of resource scarcity, especially soil erosion. In classical view, prices of the goods (*i.e.* exchange value) are taken in terms of labour input. In neo-classical economic paradigms it is relative price. Marx on the other hand endorses the value in terms of social relationship, which are historically determined. Marxian interpretation and remarks on natural resources scarcity adds to a new dimension to the whole issue, which deserve a cautious analysis in the present context.

Marshall, Hotelling and Pigou make the problem more vivid for their fellow professionals. They developed a framework where economic analysis of soil erosion seems to be feasible. They also paved the way for interventionists like Pigou, Mishan, Coase and Boulding who advocated for the strong role of state. Ronald Coase's position somehow is different who does not want state to do more than allocating the property rights for common degrading land resources. It is interesting to observe that the main culprit of land degradation is the phenomena of 'market failure' but these interventionists along with others seek remedy in market based instruments. It is also noteworthy that these interventionist's arguments were primarily focused to the problem of pollution and industrial smokes.

Soil is a complete ecosystem in itself. Development of the disciplines of ecology in the latter half of 19th century and subsequent American Conservation movement raised the issue of soil erosion. During 1940s, 1950s and 1960s there was a great leap forward in soil science. Technology and imperatives for measurement of soil loss got crystallised. This development got proper place in US Presidential Commission on Material Policy, Club of Rome, and Conferences on Environment and Development in 1987 and 1992 among others. Published reports of these efforts addressed the issue of soil erosion explicitly. They also advocated that for soil conservation economic arguments are needed and it has a wider appeal.

Thus the concern for soil erosion gradually accentuated in last 200 years. Approaches of economists to the problem have not been coherent, focused and consistent. In this context, it is not surprising that in spite of considerable advancement in research pertaining to the management of soil, it has continuously been eroded possibly because its social and economic causes hardly got highlighted. When we consider the world and the Indian scenario of soil erosion, the magnitude and dimension of soil erosion seems to be alarming.

Global degradation of soil in the form of soil loss, salinization, soil fertility decline, waterlogging etc. is quite significant. Around 17 per cent of domesticated land is degraded from light to extreme scale. Incidence of this degradation is greater in case of Central America and Africa followed by Asia. River-borne soil sediment is a good approximation of soil erosion. Here the Ganges-Brahmputra rivers alone dump around 3 billion tonnes of soil in the ocean every year.

Country-wise analysis shows the same trend. Around 42 per cent of US cropland are degraded, former USSR, China and other third world nations are facing the problem in the same severity. India's half-geographical region is facing the problem of land degradation. It is ironical that an agriculture dependent country is loosing its most precious natural resource base. India's humid slopelands (upland) primarily faces the problem of water erosion while wind erosion is causing the maximum damage in the arid and semi-arid regions. This is in spite of the fact that India is among the few countries of the world where an organised effort for estimation of soil erosion and conservation has been in action since the early 1950s.

However, estimates of soil erosion and corresponding generation of database is an area full of controversy and subjectivity. Individual researchers and government bodies of different countries come out with some baseline data on soil degradation but their approach, methodology and understanding of the problem widely vary. The UN Agency, *e.g.* UNEP and FAO conduct study in collaboration with the local experts and indigenous institutes and have provided some meaningful quantified information on global as well as individual country-level following sophisticated techniques like Geographical Information System (GIS) and remote sensing.

Natural factors like geological, geomorphological and phsiological directly cause soil erosion but for a relevant time horizon, it is human-induced force which cause the erosion most. In last 200 years global explosion of population and subsequent growth in demand for food and other developmental requirements have proliferated limitlessly. Woodcutting, cropping on the slopes and other marginal lands and overgrazing have put cumulative effect on soil erosion. Size of the land holdings and lack of permanent tenureship has further aggravated the problem.

In India the population boom, deforestation, faulty implementation of land reform, growing bovine population and marginalisation of sizeable fraction of population have been responsible for soil erosion in different parts of the country.

The impact and implications of soil erosion are purely economic in nature. The erosion of soil reduces the productivity of crops (on-site) and inflict heavy cost on the people through floods, loss in power generation and hampering navigational efficiency (off-site). Generally on-site impacts are taken into account but other direct impacts are equally important and needs serious attention. On-site impacts in terms of productivity decline are easy to evaluate but are very intriguing because the relationship of soil erosion and plant's productivity is yet not clear and precise. The offside impacts of soil erosion can be calculated easily but the source-wise specification is difficult. On *a priori* basis a definite relationship between off-site impact and on-site impacts of soil erosion does not exist. However, for a large number of cases it has empirically been found that in developed countries, off-site costs of erosion have often been assessed as substantially higher than on-site costs but in case of developing countries, the reverse may be true.

Economic valuation of soil erosion in particular and valuation of externalities in general command a special place in the multidisciplinary science of ecological economics but in this direction its evolution is in nascent stage. It derives its legitimacy from the seminal works of mainstream economists of 1950s, 1960s and 1970s. Economic valuation of the on-site and off-site impact of soil erosion rely upon techniques like defensive expenditure, replacement cost, productivity cost, property value and restotration cost. Each method gives different result but on a priori basis no one commands preference over none. Choice of method depends upon the data availability and researcher's objective.

Recently a large number of valuation of costs due to soil erosion have been undertaken by the multilateral institutions like the World Bank, FAO etc. for various African and Asian nations. Individual researchers have also attempted to evaluate the damage. Most of them use productivity loss or replacement cost approach and shows a very high monetised value for the damages inflicted. These estimations should serve the purpose of facilitating benefit-cost analysis of soil conservation programme only. Because these estimates are based on heroic assumptions and simplified version of complex ecological relationship borrowed from different natural context. Valuation of soil erosion needs indigenous and basic scientific research in tropical developing countries like India.

The analysis of soil erosion in the upland region of Dehradun valley (Doon valley) more or less resembles the international and national experiences as far as its causes are concerned. Around 2.09 lakh ha area of Doon valley has been experiencing very severe soil erosion. The period of analysis 1970-94 shows that soil loss in the valley has been much over to 100 t/ha/year. The loss of topsoil was calculated on the basis of Universal Soil Loss Equation (USLE) and for this, different parameters of USLE were also estimated for the valley. The data on land use pattern have been derived from different imageries for the period 1970-94 instead of conventional record of district statistical officer. Ideally, the analysis could have gone much beyond 1970 in the past because time-series analysis of soil loss should be based on at least 40-50 years of time period. But absence of reliable data for the valley prevented us from doing so.

Causes for the erosion in the valley are growth of population, deforestation, overgrazing, hill cultivation, unscientific mining activities and indifference of farmers towards conservation practices. Out of these factors, contribution of hill cultivation seems to be significant for soil erosion in the valley. This has precipitated due to the encroachment of fragile land as a result of population pressure.

An elaborate attempt was made to estimate the cost of lost soil in the valley for the latest year of 1994. The cost estimation was based on the production value and resource value approach. The annual (one time) cost of soil erosion based on loss in productivity lies between Rs. 1,715/ha and Rs. 3,427/ha at 1994 price while the cost on the basis of resource value in terms of lost nutrients, comes out as Rs. 21,583/ha at current price in 1994. Since the soil loss is not a one-time phenomenon, cost of accumulated loss for 20 years has been estimated. In all situations, resource value exceeds the production value of soil erosion. This confirms to other similar studies. The resource value comprehensively measures the cost and presumes all nutrients loss as the cost while plant may not use all the nutrients available in the soil and hence productivity of crops may not register all the loss due to erosion. However, as the time-horizon increases from annual to 20 years the divergence between the two approaches reduces. This may be explained in terms of the fact that as time travels in future, the productivity of crops gets hampered at accelerated rate but the rate of loss in soil nutrients occurs at the constant rate.

The cost of erosion proves to be much higher than the benefits arising out of conservation measures in the valley. The conservation measures have sought to be based on maximum possible efficiency in checking the erosion. Conservation measure devised for severely and very severely eroded area (around 36 per cent of total area of the valley) is the mix of mechanical and biological measures. Then initial cost for these measures meant to check 55 per cent of the soil loss in the valley was estimated around Rs. 85.78 crore at 1998 price, which comes at Rs. 69.48 crore at 1994 price after its deflation with wage index factor. On the basis of cost-benefit analysis, conservation measures seem to be economically efficient. Both appraisal criteria of net present value (NPV) and benefit cost ratio (BCR) conservation is adequately justified. But these encouraging mandates do not seem to be operative in the valley. In fact the reality is just contrary to this favourable signal about the feasibility of the conservation measures in Doon valley where conservation practice is either absent or is followed on a very limited scale. The reasons for this are many but primarily lie in the myopic vision of the farmers arising out of information failure. Sometimes reluctance to adopt conservation measures also emanates from the land tenureship on ad hoc basis. Conventional outlook of the farmers completes the list of reasons.

The cost of erosion in a small area of Doon valley is very high, for the country as a whole it would be astronomical. In contrast, the cost of conservation for the same might also be much lower. All appraisal criterion throw a favourable signal for conservation. But this cannot be translated into practice unless a concerted effort is made on the behalf of government. The magnitude of investible funds and time-horizon of conservation programme can only be generated and envisioned by the government. The individual farmers are constrained by the lack of information and vision. This supports the view of interventionist economist who sees a strong role for the government in management of natural resources including soil. Thus, neither Malthusian absolute limit to the quality of land seems to be in operation nor Ricardian scarcity price takes care of the erosion of soil compelling farmers to go for adequate conservation measures. And the case of soil conservation is like any other natural resource where a strong role of state is recommended to manage their depletion and degradation.

APPENDICES

APPENDIX I

PHYSIOGRAPHY, LANDUSE, SOILS, USLE'S FACTORS; SOIL LOSS ESTIMATES AND SOIL EROISON HAZARD CLASSES OF SOILSCAPE UNITS OF DOON VALLEY[A]

Mapping	Physiography	Landuse/land	Dominant soils erosion cover	Unit R	Soil erosion K	LS	Soil erosion C	P	Loss (t/ha/yr)	Class
				USLE Factor						
H—Hills (30-50 per cent slope)										
1	2	3	4	5	6	7	8	9	10	11
H11	Hills	Dense forest	[b]C.L. Typic Hapludolls	500	0.38	7.1	0.005	0.5	3.37	Slight
H12	Hills	Sparse forest	[b]L.S. Typic Udorthents	500	0.39	7.1	0.050	0.5	34.61	Severe
H13	Hills	Old terrace Cultivation/barren	L.S. Typic Udorthents	500	0.39	7.0	0.700	0.8	764.40	V. Severe
M—Mountains (30-50 per cent slope)										
M11	Mountains	Dense forests	L.S. Typic Hapludolls L.S. Dystric Eutrochrepts	500	0.39	6.00	0.005	0.5	2.93	Nil-Slight
M12	Mountains	Barren/scrub	L.S. Typic Udorthents L.S. Lithic Hapludalfs	500	0.37	6.00	0.700	0.8	621.0	V. Severe

Appendix I (*Contd.*)

1	2	3	4	5	6	7	8	9	10	11
M21	Mountains (8-10% slope)	Open forest	L.S. Typic Hapludolls L.S. Typic Eutrochrepts	500	0.40	0.42	0.600	0.7	35.28	Severe
M22	Mountains	Terrace Cultivation	L.S. Typic Udorthents L.S. Lithic Eutrochrepts	500	0.40	6.00	0.600	0.8	576.00	V.Severe
S—Siwalik Hills (10-30 per cent slope)										
S1	Northern Aspect	Dense forest	L.S. Typic Hapludolls L.S. Typic Eutrochrepts	500	0.39	7.00	0.005	0.8	5.46	Slight-Moderate
S2	Top	Open Forest	Fragmental/Stony Typic Udorthents	500	0.39	7.00	0.050	0.8	54.60	V. Severe
P—Piedmont										
P1—Upper Piedmont (4-7 per cent slope)										
P11	U. Piedmont	Dense forest	L.S. Dystric Eutrochrepts C.L. Dystric Eutrochrepts	1066	0.33	1.00	0.004	0.5	0.7	Nil-Slight
P12	U. Piedmont	Moderately dense forest	L.S. Typic Eutrochrepts C.L. Dystric Eutrochrepts	1066	0.35	1.00	0.006	0.5	1.1	Nil-Slight
P13	U. Piedmont	Cultivation (low intensity)	F.L. Eutrochrepts F.L. Mollic Ha	1066	0.40	0.24	0.400	0.6	24.56	Mo-derate

Appendix I (*Contd.*)

1	2	3	4	5	6	7	8	9	10	11
P2—Lower Piedmonts (2-3 per cent slope)										
P21	L. Piedmont	Sparse forests	F.L. Dystric Hapludalfs L.S. Dystric Eutrochrepts	1066	0.39	0.60	0.200	0.5	24.94	Moderate
P22	L. Piedmont	Degraded forests	F.L. Mollic Hapludalfs F.L. Dystric Eutrochrepts	1066	0.40	0.60	0.005	0.5	0.7	Nil-Slight
U—Uplifted Terraces										
U1	Top (2-5 % slope)	Medium intensity cultivation	F.L. Mollic Hapludalfs F.L. Dystric Eutrochrepts	500	0.39	3.0	0.050	0.8	23.4	Moderate
U2	Side Slopes (30-50 % slope)	Sparse forest/ Scrub	L.S. Typic Hapludalfs F.L. typic Eutrochrepts	500	0.39	7.0	0.050	0.7	47.78	Severe
T1—Terraces of Song and Asan (1-2 per cent slope)										
T11	Terrace	Moderately Dense forest	C.L. Aquic Eutrochrepts L.S. Typic Eutrochrepts	1066	0.50	0.14	0.006	0.5	0.22	Nil-slight
T12	Terrace	Medium to high Intensity cultivation	F.L. Mollic Hapludalfs F.L. Dystric Eutrochrepts	1066	0.52	0.18	0.450	0.8	35.92	Severe
T13	Terrace	High intensity cultivation	F.L. Typic Hapludolls L.S. Typic Eutrochrepts	1066	0.52	0.18	0.510	0.80	40.70	Severe

Appendix I (*Contd.*)

1	2	3	4	5	6	7	8	9	10	11
T2—Terraces of Ganga (1-2 per cent slope)										
T21	Terrace	Low intensity cultivation	F.L. Typic Haplaquents Typic Udipsamments	1066	0.50	0.14	0.006	0.5	0.22	Nil-slight
T22	Terrace	Medium to high Intensity cultivation	C.L. Mollic Haplaquents F.L. Typic Haplaquents	1066	0.52	0.14	0.050	0.5	1.94	Nil-slight
T23	Terrace	High intensity cultivation	C.L. Mollic Hapludalfs F.L. Typic Eutrochrepts	1066	0.52	0.18	0.050	0.5	2.49	Nil-slight
T3—Terraces of Yamuna (1-2 per cent slope)										
T31	Terrace	Low intensity cultivation	F.L. Aquic Eutrochrepts L.S. Typic Eutrochrepts	1066	0.50	0.14	0.005	0.5	0.19	Nil-slight
T32	Terrace	Medium to high intensity cultivation	F.L. Aquic Eutrochrepts F.L. Typic Eutrochrepts	1066	0.50	0.14	0.050	0.5	1.86	Nil-slight
T33	Terrace	High intensity cultivation	F.L. Typic Hapludalfs F.L. Typic Hapludolls	1066	0.52	0.14	0.050	0.8	3.10	Nil-slight

Appendix I (*Contd.*)

1	2	3	4	5	6	7	8	9	10	11
FP—Flood Plains (0-1 per cent slope)										
FP1	Flood Plain	Moderately dense forest	L.S. Dystric Eutrochrepts	1066	0.60	0.14	0.006	0.5	0.22	Nil-slight
FP2	Flood Plain	Sparse forest	L.S. Udifluvents	1066	0.61	0.14	0.050	0.5	2.27	Slight
FP3	Flood Plain	Low intensity cultivation	L.S. Dystric Eutrochrepts	1066	0.60	0.14	0.550	0.8	33.40	Severe

[b]C.L. - Coarse loamy; L.S. - Loamy skeletal; F.L. - Fine loamy; +Soil erosion hazard classes: Nil to slight (Soil loss <5 t/ha/yr); Slight to moderate (Soil loss - 5-10 t/ha/yr); Moderate (Soil loss - 11-25 t/ha/yr); Severe (Soil loss - 26-50 t/ha/yr); V. Severe (Soil loss - 50 t/ha/yr)

[a]*Source*: Centre for Inter-disciplinary Studies of Mountain & Hill Environment; CISMHE, 1994-1997.

APPENDIX II

CALCULATION OF DEFLATION COEFFICIENT FOR REAL WAGE IN UTTARANCHAL OVER 1993-94 TO 1997-98

Table A: Annual Average of Agricultural Wage Index of Uttaranchal (Base Agricultural year 1970-71)

Period	*Index*
1980-91	226.0
1981-82	261.7
1982-83	300.4
1983-84	331.7
1984-85	363.0
1985-86	402.6
1986-87	450.3
1987-88	492.3
1988-89	540.5
1989-90	620.3
1990-91	790.6
1991-92	850.6
1992-93	946.2

Source: Sankhikiya Diary 1997, Economic and Statistical Division, State Planning Institute, UP.

Table B: Annual Percentage Change in Wages for Agricultural Labour in Uttaranchal

Year	*Per cent change for agricultural year over previous year*
1992-93	(+) 7.56
1993-94	(-) 6.77
1994-95	(-) 2.31
1995-96	(+) 14.78
1996-97	(-) 6.37
1997-98	(+) 18.29

Source: Labour Statistics (pp. 146), The Indian Journal of Labour Economics January-March 1999.

On the basis of Table (A) and Table (B) the wage index for agricultural labour can be updated in 1997-98. Table (C) provides the index.

Table C: Annual Average of Agriculture Wage Index of Uttaranchal

Year	*Wage Index*
1993-94	882.14
1994-95	861.76
1995-96	982.41
1996-97	919.83
1997-98	1,080.07

Therefore the deflation coefficient of 1997-98 into 1993-94 = 882.14/ 1,088.07 = 0.81.

References

Agnihotri, R.C., L.S. Bhushan and S.P. Singh (1998). Productivity Loss due to Soil Erosion and its restoration by Conservation Practices in L.S. Bhushan *et al.* (eds.) *Soil and Water Conservation Challenges and Opportunities*, Indian Association of Soil and Water Conservation, Dehradun.

Aina, P.O., *et al.* (1982). "Effective Conservation Farming Systems for the Humid Tropics", in R. Lal (ed.) *American Society of Agronomy, Soil Erosion Conservation in the Tropics*, ASA Special Publication, No. 43, 1982.

Arrow, K., R. Solow, P.R. Portney, E.E. Leamer, R. Radner and H. Schuman (1993). *Report of the NOAA Panel on contingent valuation*, Federal Register, 15 January, 58 (10).

Bali, J.S. (1975). Address delivered at *Indian Photo-interpretation Institute*, Dehradun, India.

Bangladesh, (1972). *Land Degradation*. Paper presented to Food and Agriculture Organisation 21st Regional Conference for Asia and the Pacific, New Delhi.

Barbier, E.B. (1980). *Economics, Natural Resource Scarcity and Development*, Earthscan, London.

Barnet, H. and Morse, C. (1963). *Scarcity and Growth: The Economics of Natural Resource Availability*, Johns Hopkins University Press, Baltimore, MD.

Bennet, H. (1955). *Elements of Soil Conservation*, McGraw-Hill.

Bergsma, J. (1986). *Soil Erosion and Conservation*, Soil Conversation Society of America, Iowa.

Bishop, J. (1990). *The Costs of Soil Erosion in Malawi*, The World Bank.

—— and Allen, J. (1989). *The On-Site Costs of Soil Erosion in Mali*, The World Bank.

Bishop, R. (1982). 'Option Value: An Exposition and Extension', *Land Economics* 58.

Biswas, A. and Tewatia, R.K. (1991). 'Nutrient balance in agro-climate regions of India—An Overview,' *Fertilizer News*, 36(6).

Blaikie, P.M. (1985). *The political economy of soil erosion in developing countries*, Longman, London.

—— and H.C. Brookfield (eds) (1987). *Land Degradation and Society*, Routledge, London.

Bojo, J. (1992). Cost-Benefit analysis of soil and water conservation projects: A review of 20 empirical studies in Tato and Hurni (eds.) *Soil Conservation for Survival*, Iowa.

—— (1996).' The Costs of Land Degradation in Sub-Saharan Africa', *Ecological Economics*, 16.

Boulding, K.E. (1966). '*The Economics of the Coming Spaceship Earth*,' in H. Jarret (ed.) *Environmental Reality in Growing Economy*, Johns Hopkins Press, Baltimore, MD.

—— (1969). 'Economics as a Moral Science,' *American Economic Review*, 59.

Brandon, C. and Hormman, K. (1995). *The Cost of Inaction: Valuing the Economy-wide Cost of Environmental Degradation in India*, Asia Environment Division, The World Bank.

Brown, L.R. and Wolf, E.C. (1984). *Soil Erosion: Quiet Crisis in the World Economy*, Worldwatch Paper 60.

Buchanan, J.M. (1987). 'The Constitution of Economic Policy,' *American Economic Review*, 77.

Cantillon, R. (1755). *Essai sur la nature du commerce en general*, H. Higgs, (ed.) Macmillan, London.

Carey, H. (1837). *Principles of Economy*, Kelly, New York.

—— (1858). *The Principles of Social Science*, Kelly, New York.

Census of India (1991), Registrar General and Census Commissioner, Government of India.

Central Soil and Water Conservation Research and Training Institute, Dehradun, Annual Reports (from 1972 to 1998).

Centre for Inter-Disciplinary Studies of Mountain and Hill Environment (CISMHE), (1996). *Carrying capacity based Developmental Planning for Doon Valley*—A Report for Ministry of Environment and Forest, Govt. of India, New Delhi, 1994.

Centre for Science and Environment (CSE) (1991). A Report on Flood in India, New Delhi.

Choudhary, M.K. and Aneja, O.R. (1991). Impact of green revolution on long-term sustainability of land and water resources in Haryana, *Indian Journal of Agricultural Economics*, 45.

Ciriacy-Wantrup, S.V. (1947). 'Capital returns from soil conservation practices', *Journal of Farm Economics*, 29.

Clark, R., H. Marthrithilaka, R. White and M. Stocking (1998). 'Economic Valuation of Soil Erosion and Conservation—A Case Study of Perawella, Sri Lanka' in H-P Blume *et. al.* (eds.) *Towards Sustainable Landuse*, ISSS, Germany.

Coase, R. (1960). 'The Problem of Social Cost,' *Journal of Law and Economics*.

Crosson, P. (1985). 'National Costs of erosion on productivity,' *American Society of Agricultural Engineers*, St. Joseph, MI.

—— (1997). 'Will erosion threaten agricultural Productivity?' *Environment*, October.

—— and Anthony, T. Stout (1983). 'Productivity Effects of Cropland Erosion in the United States', Washington, D.C. *Resources for the Future*.

Dasgupta, P. (1993). *An Enquiry Into Well-Being and Destitution*, OUP.

De Janvry, A. (1981). *The Agrarian Question and Reformism in Latin America*, Baltimore, Johns Hopkins University Press.

Department of Forest, UP (1998). *Rehabilitation of Spoiled Land*, Dehradun.

Dhandapani, K.R. (1996). Economic Evaluation of Soil and Water Conservation Measures in Watersheds, *Indian Journal of Soil and Water Conservation*, vol. 40.

Dhruva Narayana, V.V. Sastry, G. and Patnaik US (1997). *Waterhead Management*, ICAR, New Delhi.

—— and Ram Babu (1983). 'Estimation of Soil Erosion in India,' *Journal of Irrigation and Drainage Engineering*, Vol. 109.

Dyson, T. (1998). *Population and Food: Global Trends and Future Prospects*, Routledge.

Eckholm, E.P. (1976). *Losing Ground*, Norton & Co.

Eckstein, O. (1961). *Water Resource Development the Economics of Project Evaluation*, Cambridge, MA.

El-Swaify, S.A. and Dangler, E.W. "Rainfall Erosion in the Tropic: A State-of-the-art", *Soil Erosion and Conservation in the*

Tropics. American Society of Agronomy and Soil Science. Special Publication No. 43.

EL-Swaify, S.A. (1993). 'Soil erosion and conservation in the humid tropics' in Pimental D. *World Soil Erosion and Conservation*, Cambridge University Press.

Elwell, H.A. and M.A Stocking (1982). 'Developing a simple yet practical method of soil loss estimation. *Tropical Agriculture (Trinidad)* 59 (I).

Ervin, D.E. (1982). Soil Erosion on Owner-Operated and rented Cropland, *Journal of Soil and Water Conservation*, Sept-October.

Estrada, R. in Southgate and Whitaker (1994). *Economic Progress and the Environment*, OUP, 1994.

Fantechi, R. Margris, N.S. (eds.), (1986). *Desertification in Europe*, Dordrecht, D. Riedel.

Fertiliser and Allied Agricultural Statistics (1994-95). Northern Region, *The Fertiliser Association of India*, New Delhi.

Fertilizer Statistics (1995-96, 1996-97). The Fertilizer Association of India, New Delhi.

Food and Agriculture Organisation, (1976). 'A Framework for Land Evaluation,' *Food And Agriculture Organisation Soils Bulletin* 32. Food And Agriculture Organisation, Rome.

Food and Agriculture Organisation, (1982). 'Population supporting capacities of lands in the Developing World,' *Food And Agriculture Organisation*, Rome.

Food and Agriculture Organisation, (1991). *The State of Food and Agriculture*, Rome.

Food and Agriculture Organisation, (1992). 'Environmental Issues in Land and Water Development,' *Food And Agriculture Organisation/RAPA,* Bangkok.

Food and Agriculture Organisation, (1992). 'Network on Erosion-induced Loss in Soil Productivity, *Food And Agriculture Organisation,* Rome.

Food and Agriculture Organisation, UNEP, *WB (1994). 'Land Degradation in South Asia: Its severity, causes and effect upon the people*, Food And Agriculture Organisation, Rome.

Food and Agriculture Organisation, *Year Book on Livestock*, 1969, 1976, 1981, 1987, 1992 and 1993.

Forestry Statistics, India, (1998). ICFRI, Dehradun.

Forrestor, J. (1971). *World Dynamics*, Wright-Allen Press, Cambridge, MA.

Freeman, A.M. IIi (1979). *The Benefits of Environmental Improvements: Theory and Practice*, Baltimore, MD.

Gardner, Gary. (1997). Pressing Global Cropland, *State of the World*, 1997.

Gazetteer of Dehradun (1911). (Reprint 1995). Natraj Publishers, Dehradun.

Georgiou, S., D. Whittington, D. Pearce and D. Moran (1998). *Economic Values and the Environment in the Developing World*, Edword Elgar.

Goldsmith, P.F. (1977). A Practical guide to the use of the Universal Soil loss equation. *Unpublished BAI, Technical Monograph*.

Gowdy, J. (1984). "Marx and Resource Scarcity: An Institutional Approach", *Journal of Economic Issues*, Vol.18, June.

Groh, F. (1994). *Economics of soil Degradation, Erosion and Conservation: A case study of Zimbabwe*, Kiel, Germany.

Gupta, G.P. (1980). 'Soil and Water Conservation in the Catchment of River Valley Projects, Status Report, *Indian Journal of Soil Conservation*, 8(1).

Gupta, P.N. (1981). *Integrated Watershed Rehabilitations and Development Project*, UP Forest Department, Lucknow, UP.

Hanemann, W.M. (1994). 'Valuing the Environment through contingent valuation', *Journal of Economic Perspective*, 8.

Heathcote, R.L. (1980). *Perception of Desertification*, Tokyo, UN University.

Hedfors, Lars (1981). 'Evaluation and Economic Appraisal of Soil Conservation in a Pilot Area', Ministry of Agriculture, *Soil and Water Conservation Branch*, Nairobi, Kenya.

Hicks, J.R. (1946). *Value and Capital*, Oxford.

Holmberg, Goran (1985). 'An Economical Evaluation of Soil Conservation in Kalia Sub-location', Kitui District, Ministry of Agriculture and Livestock Development, *Soil and Water Conservation Division, Nairobi*, Kenya.

Hotelling, H. (1931). "The Economics of Exhaustible Resources", *Journal of Political Economy*, 39: 137-175.

—— (1947). Quoted in V.K. Smith (1996) (ed.) *Estimating Economic Values for Nature*, Edward Elgar.

Hudson, N.W (1971). *Soil Conservation*. Batsford. London.

Hufschmidt, M.M., D.E. James, A.D. Meister, B.T. Bower and J.A. Dixion (1983). *Environment Natural Systems and Development: An Economic Valuation Guide*, Baltimore, Md.: Johns Hopkins University Press.

India (1998). *A Reference Manual*, Ministry of Information and Broadcasting, Government of India.

Jacks, G.V. and Whyte, R.O. (undated). *The Rape of the Earth: A World Survey of Soil Erosion*, Faber and Faber Ltd., London.

James, D. (1994). *The Application of Economic Techniques in Environmental Impact Assessment*, Kluwar Academic Publishers.

Karl Marx, (1964). "Critique of the Gotha Programme" in David Fernabch (ed.) *Karl Marx. The First International and After, Political Writings*, Vol. III, Penguin Books, Harmondsworth, England.

—— (1976). *Capital: A Critique of Political Economy*, Vol.I (Penguin Books: Harmondsworth, England.

Khosoo, T.N. and Tejwani, K.G. Soil Erosion and Conservation in India in David Pimentel (ed.) (1993) *World Soil Erosion and Conservation*, Cambridge University Press.

Kim, S.H. and J.A. Dixon (1986). 'Economic valuation of Environmental Quality Assessment of Upland Agricultural Projects in Korea in Dixm and Hufschmidt' (eds.) *Economic Valuation Techniques for the Environment: A Cast Study Workbook*, John Hopkins University Press, Baltimore.

Kneese, A.V. (1997). 'Natural Resource Economics', *In Population, Resources and Environment: Science, Ideology and Intellectual Tradition*, Washington DC.

Knight, F.H. (1924). 'Some Fallacies in the Interpretation of Social Cost', *Quarterly Journal of Economics*, 38.

Krutilla, J.V. (1967). 'Conservation Reconsidered'. *American Economic Review*, 47.

Kula, E. (1998), *History of Environmental Economic Thought*, Routledge, London.

Labour Statistics (1999), *The Indian Journal of Labour Economics*, Vol.42 (pp.146).

Lal, R. (1981) *Soil Erosion Problems on Alfisols in Western Nigeria*, VI, Effects of Erosion on Experimental Plots, Geoderma, 25: 215-230.

—— (ed.) (1994) *Soil Erosion Research Methods*, St. Lucie Press.

Landuse Survey Directorate, Forest Department Uttar Pradesh State (1992). *A Study of Land, Vegetation and Water Resources of the Himalayan Region of Uttar Pradesh.*

Larson, W.E. *et al.* (1985). 'Loss in long-term productivity from soil erosion in the US. In El-sweify, S.A. *et al. Soil Erosion and Conservation.*

Longdale, G.W. and Shrader, W.D. (1981). 'Soil erosion effects on soil productivity of cultivated cropland.' In Determinants of Soil Loss Tolerance, *American Society of Agromony*, Madison, Wisconsin.

Lyles, Leon (1975). Possible effects of wind erosion on soil productivity, *Journal of Soil and Water Conservation*, 30 (6).

Magrath, W. and Arens, P. (1989). *The Costs of Soil Erosion in Java: A Natural Resource Accounting Approach*, The World Bank.

Magrath, W.B. (1980). '*Economic Analysis of Soil Conservation Technologies'*, Environment Department Division Working Paper, 1989-94, The World Bank.

Malthus, A. (1978). *An Essay on the Principle of Population as it Affects the Future Improvement of Society*, 3rd edition, Ward Lock, London.

Markandya, A. and D.W. Pearce (1988). *Environmental Considerations and the choice of Discount Rate in Developing Countries*, Environmental Department Working Paper No. 3, The World Bank.

Marsh, G.P. (1865). *Man and Nature: Or Physical Geography is Modified by Human Action*, Charles Scribber, New York.

Marshall, A. (1890). *Principles of Economics*, Macmillan, London.

McKenzie, C. (1994). *Degradation of Arable Land Resources: Policy options and considerations within the context of Rural Restructuring in South Africa*, LAPC Workshop, Johanesberg, South Africa.

Meadows, D.H. *et al.* (1972). *The Limits to Growth*, Pan Books, London.

Mill, J.S. (1848). *Principles of Political Economy*, Appleton, New York.

Miller, M.F. (1980). *Soil Degradation: Effects on Agricultural Productivity, National Agricultural Study*, Washington DC.

Ministry of Agriculture (1985). *Indian Agriculture in Brief*, 20th edition, Govt. of India.

Ministry of Agriculture, (1992). *Indian Agriculture in Brief*, 24th Edition.

Mishan, E.J. (1967). *The Costs of Economic Growth*, Penguin, London.

—— (1977). *The Economic Growth Debate*, Allen and Unwin, London.

MOA (1968). Report of the Working Group for formulation of *The Fourth Five Year Plan Proposal for Land and Water Development*, Ministry of Agriculture, Government of India, New Delhi.

MOA (1985). *Indian Agriculture in Brief*, 20th edition, MOA, Govt. of India, New Delhi.

National Bureau of Soil Survey and Landuse Planning (1990). *Productivity of Crops and Soil Degradaton in India*, Nagpur.

Nowak, P. (1987). 'Implementation of Soil and Water Conservation' in Hallback *et al.* Making Soil and Water Conservation Work, Soil and Water Conservation Society, Iowa.

Nowak, P.J. (1988). The Costs of Excessive Soil Erosion, *Journal of Soil and Water Conservation.*

Pagiola, Stefano, (1992). 'Estimating the Yield Effects of Soil Degradation', *Journal of Soil and Water Conservation*, 47.

Pandey, A.N., Pathak, P.C., and J.S. Singh. (1984), 'Water, sediment and nutrient movement in forested and non-forested catchments in Kumaon Himalaya, *Forest Ecology and Management*, 7, 1929.

Parikh, J.K. and Parik, K.S. (1997). *Accounting and Valuation of Environment*, The UN, NY.

Pesek, J. (1980). Unified resource conservation in agriculture: Roles in agronomy. *Agronomy Journal*, 72.

Pesner, J. and McPherson, M. "Agriculture on the Steep Slopes of Tropical America", *World Development* 10:2 (1982).

Pigou, A.C. (1929). *Economics of Welfare*, Macmillan, London.

Pimentel, D. (1993). *World Soil Erosion and Conservation*, Cambridge University Press.

—— *et al.* (1976). 'Land Degradation: Effects and Food and Energy Resources,' *Science*, 194.

—— *et al.* (1987). 'World Agricultural and Soil Erosion,' *Bioscience*, 37(4).

Poletayer, P. and Yashukova, S. (1978), "Environmental Protection and Agricultural Production". *Ekonomika Selskogo Khozyaystta*, Moscow.

Postel, Sandra, (1989). *Halting and Degradation*, State of the World.

Pradhan, S.B. (1992). *Status of Fertilizer in the Asia Pacific Region*, Paper presented to FADINAP Seminar, Chiang Mai, Thailand.

Ramsey, F. (1928). A Mathematical Theory of Savings, *Economic Journal.*

Ravishakar, H.M., S.K. Srivastava, S.K. Saha, Pramod Kumar and Jitendra Prasad (1994). 'Watershed Priotization through the Universal Soil Loss Equation Using Digital Satellite Data and an Integrated approval' *Asian Pacific Remote Sensing Journal*, Vol.6.

Ricardo, D. (1817). *Principles of Political Economy and Taxation*, Pelican Books, London.

Ridker, R.G. (1967). *Economic Costs of Air Pollution*, New York: Praeger.

Rosenberry, Paul *et al.* (1980). Predicting Effects of Soil Depletion from Erosion, *Journal of Soil and Water Conservation.*

Saha, S.K., J. Bhattacharjee, C. Lalengzuva and L.M. Pande (1992). 'Prioritisation of subwatersheds based on Erosional Soil Loss Estimates—A Case Study of Part of Song River Watershed, Doon Valley,' Using Digital Satellite Data in *Proceedings of National Symposium for Sustainable Development.*

Samuelson, P.A. (1983). *Foundation of Economic Analysis*, Enlarged edition, Harvard University Press.

Sankhikiya Patrika, District Dehradun, Economic and Statistical Planning Institute, UP (From year 1972 to 1997).

Saunders, I. and Young, A. (1983). Rates of Surface Processes on slopes, slope retreat and denudation. *Earth Surface Processes and Land Farm*, 8.

Schertz, D.L. (1983). The basis for soil loss tolerances, *Journal of Soil and Water Conservation*, 38.

Schumpeter, J. (1954). History of Economic Analysis, Allen & Unwin, London.

Singh, Gurmel, Ram Babu, Pratap Narayan, L.S. Bhushan and I.P. Abrol (1992). 'Soil Erosion rates in India,' *Journal of Soil and Water Conservation*. 47.1

Sinha, B. (1984). *Role of Watershed Management in Water Resources Development Planning*, East-West Centre, Honolulu, Hawaii.

Smith, A. (1976). *An Inquiry into the Nature and Causes of the Wealth of Nations*, Routledge, London.

Smith, V.K. (1992). 'Environmental Costing for Agriculture: Will it Be Standard Fare in the Farm Bill of 2000?, *American Journal of Agricultural Economics*, 74.

—— (1993). 'Non-market Valuation of Environmental Resources: An Interpretive Appraisal', *Land Economics*, 67.

Southgate, D. (1988). *The Economics of Land Degradation in the Third World*, *Environment Department Working Paper No. 2*, The World Bank.

—— (1999). 'The causes for land degradation along' "Spontaneously" Expanding Agricultural Frontiers in the Third World, *Land Economics*, Vol.66.

—— and Whitaker, M. (1994). *Economic Progress and The Environment*, Oxford University Press.

Stocking, M. (1986). The Cost of Social Erosion in Zimbabwe in Terms of The Loss of Three Major Nutrients, *Food And Agriculture Organisation*, Rome.

—— (1988). 'Socioeconomic of Soil Conservation in Developing Countries,' *Journal of Soil and Water Conservation*, Sept.-Oct., 1988.

Stout, B.A. (1965). *Soil erosion by water, some measures for its control on cultivated lands*, Food And Agriculture Organisation, Rome.

Tandon, H.L.S. (1992). *Assessment of Soil Nutrient Depletion*, Paper presented to FADINAP Seminar Chiang Mai, Thailand.

Tata Energy Research Institute (TERI) (1998). *Green India 2047: Looking Back to Think Ahead*, New Delhi.

Taussing, M. (1978). "Peasant Economics and the Development of Peasant Agriculture in the Cauca Valley, Columbia", *Latin American Perspective 18:5.*

Turner II, B.L. *et al.* (1990). *The Earth as Transformed by Human Action*, Cambridge University Press.

Turvey, R. (1963), "On Divergence Between Social Cost and Private Cost", *Economica* 30 (119).

UN (1977). *Desertification: Its Causes and Consequences*, Oxford, Pergman.

UNEP, *Environmental Data Report, 1993-1994*, Basil Blackwell.

USAID, (1979). *Environmental and Natural Resource Management in Developing Countries: A Report to Congress*, Washington DC.

USDA (1982). Soil Conservation Service, Basic Statistics and Preliminary National Resource Inventory.

USDA Economic Research Service, (1983). "World Agriculture Outlook and Situation", Washington DC, June.

USDA, Economic Research Service, (1981). Economic Indicators of the Farm Sector: Production and Efficiency Statistics, Washington DC.

Veloz, A., D. Southgatge, Fred Hitzhusen and R. Macgregor (1985). 'The Economics of Erosion Control in a Subtropical Watershed: A Dominican Case', *Land Economics*, 61.

Vohra, B.B. (1982). *Land and Water Management Problems in India*, Training Volume 8 (New Delhi, Ministry of Home Affairs.

Walling, D.E. (1987). "Rainfall, Runoff and Erosion of the Land: A Global View" in K.J. Gregory (ed.). *Energetics of Physical Environment*, John Wiley and Sons, New York.

WCED, (1987). *Our Common Future*, OUP.

Wild, A. (1993). *Soils and the Environment: An Introduction*, Cambridge University Press.

Williams, J.R. (1985). 'The Physical Components of the EPIC mode' in El Swaify, *Soil Erosion and Conservation.*

Wischmeier, W.H. and Smith D.D. (1965). 'Predicting rainfall erosion losses from cropland east of the Rocky Mountains, *Handbook No. 282,* USA, Washington DC.

—— (1978). *Predicting rainfall erosion losses*. Washington DC: US Department of Agriculture. (Agricultural handbook No. 537).

Woodsworth in L.R. Brown and E.C. Wolf. Soil Erosion: Quiet Crisis in the World Economy, 1984. *Worldwatch Paper 60*.

WRI, 1998-99. *World Resources: A Guide to the Global Environment*.

Zachar, D. (1982). *Soil Erosion*, Elsevier Scientific Publications.

Index